Between Probability and Certainty

Between Probability and Certainty

What Justifies Belief

Martin Smith

OXFORD
UNIVERSITY PRESS

OXFORD
UNIVERSITY PRESS

Great Clarendon Street, Oxford, OX2 6DP,
United Kingdom

Oxford University Press is a department of the University of Oxford. It furthers the University's objective of excellence in research, scholarship, and education by publishing worldwide. Oxford is a registered trade mark of Oxford University Press in the UK and in certain other countries

First Edition published in 2016

Impression: 1

Published in the United States of America by Oxford University Press
198 Madison Avenue, New York, NY 10016, United States of America

British Library Cataloguing in Publication Data
Data available

Library of Congress Control Number: 2015943402

ISBN 978–0–19–875533–3

Printed in Great Britain by
Clays Ltd, St Ives plc

This book is dedicated to Leigh Smith (1945–2014)

Contents

Acknowledgements

The ideas at the heart of this book originated while I was a PhD student at the Australian National University and reading, at much the same time, David Lewis on counterfactuals, Paul Pietroski and Georges Rey on ceteris paribus laws and Roy Bhaskar on tendencies. After completing my PhD in 2005, I developed these ideas in a series of papers—principally 'Ceteris paribus conditionals and comparative normalcy' (*Journal of Philosophical Logic*, 2007), 'What else justification could be' (*Noûs*, 2010), 'A generalised lottery paradox for infinite probability spaces' (*British Journal for the Philosophy of Science*, 2010) and 'Knowledge, justification and normative coincidence' (*Philosophy and Phenomenological Research*, 2014). The ambition to try and give this material a proper book length treatment only became serious in 2013, and the book was finally completed during a period of research leave funded by the Arts and Humanities Research Council from September 2014 to May 2015.

There are many who have helped me in one way or another with the ideas in this book and to whom I owe thanks: Renée Bleau, Michael Blome-Tillmann, Peter Brössel, Campbell Brown, Adam Carter, Jake Chandler, Martin Davies, Dylan Dodd, Igor Douven, Anna-Maria Eder, Andy Egan, David Enoch, Talia Fisher, Brian Garrett, Benoit Gaultier, Emma Gordon, Patrick Greenough, Ulrike Hahn, Allan Hazlett, Carrie Ichikawa Jenkins, Ben Jarvis, Jonathan Jenkins Ichikawa, Jesper Kallestrup, Cristoph Kelp, Gary Kemp, Stephan Leuenberger, Clayton Littlejohn, Conor McHugh, Neil Mehta, Bob Meyer, Peter Milne, Velislava Mitova, Luca Moretti, Ram Neta, Paul Pietroski, Duncan Pritchard, James Pryor, Peter Roeper, Jan-Willem Romeijn, Daniele Sgaravatti, Stewart Shapiro, Levi Spectre, Wolfgang Spohn, Ernest Sosa, Asbjørn Steglich-Petersen, Allard Tamminga, Claudine Tiercelin, Nick Treanor, Jonathan Way, Brian Weatherson, Ralph Wedgwood, Alan Weir, Crispin Wright, and Elia Zardini. Lists of this kind have become conventional, but it is difficult to compile them in a satisfactory way, particularly when a book has been many years in development. My apologies go to anyone that I have inadvertently left off. Particular thanks are due to David Atkinson, Philip Ebert, Aidan McGlynn, Jeanne Peijnenburg, and Adam Rieger, each of whom generously read a near complete draft of the book and provided extensive comments, and to Kenny Easwaran and Matthew McGrath, the two readers for Oxford University Press, whose thorough, thoughtful reports led to a number of substantial improvements.

Material related to this book has been presented a number of times: at the Australasian Association of Logic Conference at the University of Western Australia in September 2005, at the University of St Andrews in October 2006 and October 2007, at the first Basic Knowledge Workshop at the University of St Andrews in November 2007, at the University of Glasgow in September 2008, January 2014, and February 2015, at the University of Stirling in October 2008, at the Arché/CSMN Graduate Conference at the University of Oslo in November 2008, at the University of Western Australia in June 2009, at the University of Aberdeen in June 2011 and March 2012, at the University of Edinburgh in January 2012 and December 2012, at Queen's University Belfast in February 2012, at the Justification, Normality and Normativity Conference at the Institut Jean Nicod, Paris in June 2012, at the Aims and Norms Workshop at the University of Southampton in June 2013, at the New Perspectives on External World Scepticism Conference at the Munich Centre for Mathematical Philosophy in June 2013, at the University of Groningen in April 2014, at the Summer Philosophy conference at Yonsei University, Seoul in June 2014, at the Norms of Reasoning Workshop at the University of Bochum in September 2014 and at the Statistical Evidence in Epistemology and the Law Workshop at the University of Glasgow in December 2014. My thanks to all of those who participated on these occasions.

I wish to express my thanks to Peter Momtchiloff and to the editorial team at Oxford University Press for their invaluable help and advice in preparing this book. Most of the material in this book is either new or substantially reworked—my thanks to Oxford University Press for permission to reprint proofs in Chapter 9 that originally appeared in the *British Journal for the Philosophy of Science*. Thanks also to Kevin Robertson for generously allowing me to reproduce his artwork on the cover. Lastly, I am very grateful to my family—my wife Meredith, my children Alexander and Madeleine, my mother Janine, and my sister Mireille. This book would never have been completed without their love and support. My final thanks go to my father Leigh. My very first philosophical conversations, and the ones that I remember most vividly, were all with him. Without his unwavering encouragement and belief in me, I would never have even imagined that philosophy was something I could pursue.

Introduction

The Risk Minimisation Conception of Justification

Some philosophers have claimed that cases involving lotteries provide vivid counterexamples to the traditional analysis of knowledge as justified, true belief (see Hawthorne 2003: 9, Pritchard 2007: 4). They reason along the following lines: suppose I hold a single ticket in a fair lottery of one million tickets. Suppose I am convinced, purely on the basis of the odds involved, that my ticket won't win. Do I *know* that my ticket won't win? Intuitively, I don't know any such thing, even if it happens to be true. Presumably, though, I have plenty of *justification for believing* that my ticket won't win—after all, given my evidence, this proposition has a 99.9999% chance of being true. How much more justification could one want? If I'm not justified in believing that my ticket won't win, then surely none of us is justified in believing much at all. Here is a case, then, in which a justified, true belief fails to qualify as knowledge.

This argument seems straightforward enough, and yet there are reasons for being uneasy about it. On reflection, lottery cases seem somehow *different* from the standard Gettier cases that are used to refute the traditional analysis of knowledge. Consider the following: I wander into a room, undergo a visual experience as of a red wall and come to believe that the wall is red. In actual fact the wall *is* red but, unbeknownst to me, it is bathed in strong red light emanating from a hidden source, such that it would have looked exactly the same to me even if it had been white. Intuitively, I do not know, in this case, that the wall is red, in spite of the fact that my belief is both justified and true.

We can observe a number of apparent differences between these two cases. In particular, while my belief in the Gettier case fails to actually qualify as knowledge, it nevertheless seems to be a good or promising *candidate* for knowledge—and would have been knowledge if only conditions in the world had been more obliging. My belief in the lottery case, however, doesn't seem to be the sort of belief that could *ever* qualify as knowledge. In the Gettier case, the problem seems to lie with the *world*—and funny, abnormal goings on therein. In the lottery case, the problem seems to lie with *me* and the way in which I form my belief (see Ryan 1996: 136, 137).

The Hawthorne/Pritchard argument betrays a commitment to a certain, quite pervasive, way of thinking about epistemic justification. The picture is something like this: for any proposition P we can always ask how *likely* it is that P is true, given present evidence. The more likely it is that P is true, the more justification one has for believing that it is. The less likely it is that P is true, the less justification one has for believing that it is. One has justification *simpliciter* for believing P when the likelihood of P is sufficiently high and the risk of ~P is correspondingly low. Call this the *risk minimisation* conception of justification.

This general sort of picture can be identified in the work of a very broad range of epistemologists. Sometimes it is made more or less explicit (see Russell 1948: chap. VI, Chisholm 1957: 28, Derksen 1978, Alston 1988, Moser 1988, Plantinga 1993: chap. 9, Fumerton 1995: 18–19, Lewis 1996: 551, Swinburne 2001: chap. 3, Swinburne 2011, Conee and Feldman 2004: n. 32, Pryor 2004: 350–1, Pryor 2005: 181, BonJour 2010, Goldman 2011: s. 16.7). More often it is left implicit, as in the above reasoning. I don't know of any detailed arguments in favour of this picture, though the following thought is admittedly quite compelling: most epistemologists are *fallibilists* of one kind or another and hold that a belief can be justified even if one doesn't have evidence that makes it certain—even if one hasn't completely eliminated all risk of error. But if justification can fall short of evidential certainty, then what else could it possibly *be* if not evidential probability or likelihood? If justification does not require the complete elimination of error risk, then what else could it possibly require if not its minimisation?

When all is said and done, I'm unsure whether I can offer adequate answers to these questions—but I will attempt, in this book, to come to a rather different way of thinking about justification. Some of the views that I'll defend—or at least take seriously—might strike some as obviously wrong. An example might be the view that we can, sometimes, be justified in believing things that are very unlikely to be true, given our evidence. I think that this may in the end be correct—and by 'justification', I don't simply mean justification that is 'practical' or 'prudential', but justification that is *genuinely epistemic*. But I will begin with ideas that are, I hope, less controversial.

It's clear that there is *something* that sets purely statistical evidence apart from evidence of other kinds. In the case described above, my evidence for the proposition that my ticket will lose the lottery is a clear example of evidence that is purely statistical in character. Here is another (based on an example due to Cohen 1977: §24): Suppose that 100 people attended a concert but only one ticket was ever sold. As such, only one person at the concert attended legitimately and the other 99 were gatecrashers. Suppose we know that Joe was one of the people who attended the concert but have no further information about him. Is Joe a gatecrasher? In one sense it has to be admitted that our evidence in favour of this proposition is very strong indeed. And yet, when pressed, most of us would, I think, be reluctant to give this evidence much weight.

Should I draw the conclusion that Joe gatecrashed the concert and treat him accordingly? Should I go about asserting that Joe is a gatecrasher—should I, for instance, inform his friends, his family, his employer? Should Joe be taken to court and appropriate punishment applied to him? Most of us would be very apprehensive about taking such steps just on the grounds that Joe attended the concert and that 99 out of 100 attendees were gatecrashers. And this is not just a philosopher's intuition—under prevailing legal practice, in a broad range of jurisdictions, statistical evidence of this kind would not be deemed sufficient for a positive finding of fact to the effect that Joe gatecrashed the concert (for some relevant references see Kaye 1982: s. I, Allensworth 2009: s. IIB). From the perspective of the risk minimisation conception, though, this apprehensiveness is puzzling. After all, there's no question that our evidence makes it *very likely* that Joe gatecrashed the concert. By believing, asserting, and acting upon this proposition, we would only be running a very small risk of error.

Perhaps when it comes to believing and asserting that Joe gatecrashed and applying appropriate sanctions to him, we wouldn't be willing to tolerate *any* risk of error, no matter how small. But this really seems not to be the case—for we would be perfectly willing to do such things on the strength of *other kinds* of non-conclusive evidence. Suppose that, instead of our having statistical evidence in favour of Joe being a gatecrasher, we have some eyewitness testimony to that effect. Suppose a witness testifies that she clearly saw Joe scaling the fence at the concert or some such.

As long as we had no reason to doubt the reliability of this testimony, we would usually be willing to take it at face value—and to repeat it and act upon it. Most of us would be quite comfortable with Joe being appropriately punished on the strength of evidence such as this. And, under prevailing legal practice, testimonial evidence of this kind, provided it is not contradicted or otherwise called into question, could constitute sufficient grounds for a finding to the effect that Joe gatecrashed the concert.

We are all perfectly aware, though, that testimony is fallible. Just because an eyewitness testifies that Joe gatecrashed the concert, this doesn't *make it certain* that he did—he may still be innocent. Witnesses are sometimes mistaken and they sometimes lie. The eyewitness testimony undoubtedly makes it *likely* that Joe gatecrashed but, plausibly, it doesn't make it quite as likely as 99%—and yet this is precisely how likely the proposition is, given the statistical evidence about which we seemed so apprehensive. By believing that Joe gatecrashed on the basis of the testimonial evidence, we would actually be running a *higher* risk of error than we would in believing this on the basis of the statistical evidence. As such, the risk minimisation conception straightforwardly predicts that the latter belief should be more justified than the former. Yet this would seem to be the very opposite of the truth.

Maybe there is nothing here that should seriously trouble a risk minimisation theorist. Maybe our judgements about the case reflect an unreasonable bias against statistical evidence and we should train ourselves to give them up. But if we were to try and devise a theory of justification that really did do justice to our judgements here, what would it look like? Apart from anything else, it would need to be a view on which justification somehow demanded *more* than probability, but *less* than certainty—a view on which no amount of purely statistical evidence could make for justification, even though something like testimonial evidence somehow could.

As I've mentioned, the testimony to the effect that Joe gatecrashed the concert still leaves open that possibility that he is innocent. This was an open possibility before we received the testimony and it remains so afterwards. What the testimony does, though, is force us to *reconceptualise* this possibility—to view it in a different sort of light. Once we've received the witness testimony, there is *something that it would take* in order for Joe to still be innocent—it would take a deceit, it would take a misperception, it would take some sort of departure from *normal circumstances*.

In contrast, if my only evidence against Joe is that he attended the concert and that 99 out of 100 attendees were gatecrashers then, while it may be unlikely that Joe is innocent, it wouldn't really take any departure from normal circumstances for this to be true. While I might, in a sense, be surprised to learn that Joe is innocent, in another sense there shouldn't be anything particularly surprising about this, given my evidence. I know that one of the attendees was innocent—and it is no more surprising that this should turn out to be Joe than anyone else. If I believe that Joe gatecrashed, on the basis of the statistical evidence, then my belief could turn out to be false without anything abnormal having transpired. If I believe that Joe gatecrashed on the basis of testimony, then it would take some abnormal circumstance to part my belief from the truth. Viewed in this way, the testimonial evidence really does seem to offer something more than the statistical evidence does.

These observations are in no way peculiar to testimonial evidence per se. Suppose *I* am an eyewitness to Joe's gatecrashing—suppose I get a clear look at him scaling the fence. Like testimonial evidence, direct perceptual evidence does more than just 'load the dice' in favour of a proposition. My perceptual experience may make it less likely that Joe is innocent, but it also makes this possibility *demand something more*—hallucination, perceptual malfunction, disguise, etc.—it makes it demand a departure from normalcy. Purely statistical evidence is distinctive for *not* having this effect.

These remarks do not, of course, amount to some new theory of justification—not yet. At present, they are perhaps little more than suggestive slogans—and they contrast sharply with my formulation of the risk minimisation conception as a relatively clear and precise thesis. My primary aim, in this book, is to try and make something more of these ideas—to build them into something that would count as a viable alternative to the risk minimisation conception.

While my way of thinking about justification may be unfamiliar, it is connected to a more familiar way of thinking about *knowledge*. Many epistemologists have been attracted to the idea that, in order for a belief to qualify as knowledge it must be *safe* from error—it is necessary that the belief could not easily have been wrong. This is often spelled out in terms of possible worlds: in order for a belief to be knowledge it must be true in all close or similar possible worlds at which it is held. In order for a belief to be justified, on my view, there is also a set of possible worlds throughout which the belief must be true wherever it is held. These are not worlds that are 'close' or 'similar' as such, but worlds that are *normal.*

One might wonder whether those who adopt a safety condition upon knowledge have some additional incentive to adopt a structurally similar condition on justification, such as the one I defend here. There may be some truth to this—but we should proceed with caution. What is clear, though it has not been widely observed, is that combining a safety condition upon knowledge with the risk minimisation conception of justification makes for a kind of *ill-fit* between the two norms. As I will be arguing, there are in fact a number of familiar ideas about knowledge that sit very uneasily alongside the risk minimisation conception of justification.

I mentioned above that the risk minimisation conception of justification is left implicit in the work of a number of epistemologists. But the acceptance of this picture is, by and large, more self-conscious in the so-called 'formal epistemology' tradition where it has been brought to the fore via the paradoxes of rational acceptability. The risk minimisation conception, as it stands, is incompatible with the principle that justification is closed under multiple premise deductive consequence—the principle according to which, if one has justification for believing each of a set of premises, and these premises together deductively entail a conclusion, then one has justification for believing the conclusion. The lottery and preface paradoxes both serve to make this incompatibility vivid.

In spite of its incompatibility with the letter of the risk minimisation conception, the closure principle is one that some epistemologists hold dear. As such, there have been a number of attempts to refine or modify the risk minimisation conception in the hope of circumventing the lottery and preface paradoxes without having to abandon multiple premise closure. Such attempts turn out, however, to be beset by purely formal difficulties. A series of impossibility results, as I shall discuss, come close to demonstrating that nothing recognisable as a refinement of the risk minimisation picture can be consistently combined with multiple premise closure.

My own theory of justification, in contrast, *is* consistent with multiple premise closure, and offers a different sort of treatment of the lottery and preface paradoxes. It's important to bear in mind that these paradoxes are not puzzles waiting to be 'solved' in such a way as to preserve *all* of our pretheoretic impressions. Any treatment of the paradoxes will have to trade off advantages against disadvantages—and my own theory of justification simply strikes this bargain in a different way to the risk minimisation conception. The lottery and preface paradoxes will both be discussed in detail, along with a number of related puzzles.

In this book, I will be presenting a range of arguments against the risk minimisation conception of justification—arguments of which this introduction has offered a brief preview. But it's not on the strength of arguments that the risk minimisation conception ever became the dominant view and, I suspect, it won't be dislodged by arguments either—or, at least, not by arguments *alone*. The risk minimisation conception is part of a complex of ideas that really do have a remarkable coherence and power—and this may go a long way towards explaining its enduring popularity. It's for this reason that the provision of some kind of alternative is of the utmost importance. My primary aim here is to develop this alternative.

1

Two Epistemic Goals

1.1. Aiming for Knowledge, Aiming for Justification

I begin with an unqualified assumption. While I'm inclined to think this assumption is true, I am unsure of how best to argue for it, so simply put it forward for now. What can be argued is that the assumption leads to some surprising consequences—consequences that turn out to be at variance with a number of prevailing views in epistemology. The assumption concerns our epistemic *goals*.

Justification and knowledge are two worthy epistemic goals to which we might aspire in forming beliefs. When we come to believe something, we would like the belief to be justified, and we would like it to be knowledge. These are separate goals insofar as we could achieve one without achieving the other. In particular, most would agree that we could achieve the former goal without achieving the latter—a belief could qualify as justified without qualifying as knowledge. If my assumption is correct, though, there is another sense in which these goals cannot be separated—they are, to use Crispin Wright's phrase, *normatively coincident* (Wright 1992: 18–19). Put simply, two goals are normatively coincident just in case there is no way to *aim* for one without automatically aiming for the other—though it may be possible to *succeed* at one without succeeding at the other.[1]

[1] For Wright, this characterisation of normative coincidence is offered as a casual gloss on a more formal definition: 'Say that two predicates *coincide* in (positive) normative force with respect to a practice just in case each is normative within the practice and reason to suppose that either predicate characterises a move is reason to suppose that the other characterises it too' (Wright 1992: 18). The specifics of this definition very much reflect the overall dialectic in which it is embedded in the opening chapter of *Truth and Objectivity*—and are less of a natural fit with my concerns here. As such, I shall make do with the former characterisation.

Suppose I'm a member of a running club, due to compete in a race. Here are two goals to which I might aspire: (1) To set a new club record. (2) To set a new personal best. Let's fill in some further details: suppose that my best times are all official club times and, as such, I can't set a new club record without also setting a new personal best. Suppose that, while I do hold some of the fastest club times, I'm aware that I don't hold the very fastest time and, as such, I could set a new personal best without setting a new club record. Goals (1) and (2) are clearly distinct—I could accomplish (2) without accomplishing (1). Given the set up, though, I could not *aim* to accomplish (2) without *aiming* to accomplish (1)—not, at any rate, by any of the ordinary ways that one might prepare for a race. Whatever things I need to do in order to try and set a new personal best (stay fit, train hard, watch my diet, etc.) these are exactly the things that I need to do in order to try and set a new club record—nothing more and nothing less.

Another goal that I could have in mind, in preparing for the race, is just to get a *respectable* time—a time that's not far below the present club average, say. Clearly, I can't set a new club record without also getting a respectable time, but I could get a respectable time without setting a new club record. It's relatively easy, though, to imagine a course of action directed at getting a respectable time that would scupper my chances of setting a new record—I could, for instance, take it deliberately easy with the training, not push myself too hard, etc. While setting a new personal best and getting a respectable time are both more modest goals than setting a new club record, the former *normatively coincides* with this goal while the latter does not.

To put it differently, the goal of setting a new club record and the goal of just getting a respectable time potentially *compete* with one another—pull me in different directions. When I'm preparing for the race, turning over various training regimens in my mind, I may well have to weigh up the relative merits of these two goals—if I take them both seriously at any rate. One thing I won't need to do though is weigh up the relative merits of setting a new club record vs setting a new personal best. I could well take a view on which of these is more important to me but, in a sense, it would be purely academic. When it comes to figuring out how to train for the race, these goals might just as well be treated as one and the same. If my starting assumption is correct, then the relationship between knowledge and justification is more like the relationship between setting a new club record and setting a new personal best—and less like the relationship between setting a new club record and getting a respectable time.

It is, admittedly, somewhat unusual to think about knowledge and justification in this way. And the claim that knowledge and justification are normatively coincident has never, as far as I'm aware, been made—at least not in so many words. Nevertheless, this claim—or something very close to it—may be implicit in some recent theories of justification. Alexander Bird (2007) and Jonathan Jenkins Ichikawa (2014) have proposed that justification can be thought of as a kind of *would-be* knowledge. On Bird's view, roughly put, if a subject with certain mental states forms a belief, then that belief will count as justified iff there is a possible world in which the subject, with the same mental states, forms a corresponding belief and the belief qualifies as knowledge. For Ichikawa, a subject's belief is justified just in case a possible counterpart of that subject, identical in relevant intrinsic respects, holds a corresponding belief that qualifies as knowledge. On these views, a belief is justified if it is knowledge or if it fails to be knowledge as a result of interference from external or extra-mental factors. In a similar vein, Steven Reynolds (2013) has proposed that justification can be understood as the *appearance* of knowledge. On Reynolds's view, roughly, a subject's belief qualifies as justified iff, during the time in which the subject acquired and maintained the belief, it appeared to the subject to be knowledge.

None of these philosophers discusses the question of normative coincidence. But, if any of these proposals is on the right track, then justification and knowledge could be expected to normatively coincide. Aiming to form a belief that *appears*, to one, to be knowledge is presumably the same thing as aiming to form a belief that really *is* knowledge. Similarly, aiming to form a belief that *would* constitute knowledge, given the right external, extra-mental conditions, is presumably the same thing as aiming to form a belief that *does* constitute knowledge. There is one sense in which all that I can really do, ultimately speaking, in aiming to achieve knowledge, is to get my own internal, mental states in order and then put my faith in external, extra-mental conditions to fall into place.

One thing that is striking about all of these theories is that they place *knowledge* squarely in the driver's seat. That is, these theories all portray knowledge as the primary epistemic goal and seek to understand justification as a kind of *approximation* of knowledge or as a *next best thing* to knowledge. But the claim that justification and knowledge are normatively coincident, in and of itself, carries no such implication. On the contrary, it is consistent with this idea that *justification* is the primary epistemic goal and that knowledge is some sort of extrapolation or projection from it. It is also consistent with this idea that neither justification nor knowledge enjoys any special priority over the other.

Apart from these proposals, there is another theory of justification that straightforwardly predicts its normative coincidence with knowledge—namely, the theory proposed by Sutton (2007) according to which justification *just is* knowledge. On Sutton's theory, a belief is justified iff it constitutes knowledge. This theory obviously guarantees that justification and knowledge are normatively coincident goals, though only at the (considerable) expense of denying that they are distinct goals at all.

I will have more to say about each of these theories along the way. Whatever we make of them, though, my sympathy for the idea that knowledge and justification are normatively coincident doesn't derive from a sympathy with any particular theory of justification. In a way, I'm simply struck by the impression that knowledge and justification can't compete with one another in the way that other goals can. When I'm preparing for a race I might well decide that I'll take it easy and aim for a respectable time, rather than a club record. But when I'm inquiring into something, trying to get at the truth, I can't decide that I'm going to 'take it easy' this time and go for justification, rather than knowledge—this doesn't seem to make sense at all. What am I to do differently? More generally, it seems that inquiring into a matter and arriving at some conclusion about it is not an activity that is governed by competing epistemic goals. When deliberating over what to believe about a given issue I might weigh up available reasons and evidence, but I don't weigh up different *epistemic goals*—or so it seems to me.

Suppose I'm about to face a driving test. Suppose I've had several driving tests in the recent past and failed every one. As I sit in the car about to turn the key in the ignition, I could imagine agonising over what I should expect or anticipate about this latest test. On the one hand, my evidence seems to suggest that I'll fail—so this might be the most realistic thing to believe. On the other hand, how can it be reasonable to go into the test already expecting to fail? Such a belief would surely increase my anxiety and may even impair my performance, so perhaps I should just suspend judgement about this or, if possible, even try to convince myself that I'll pass, etc. This is the clearest sort of example I can think of in which different goals putatively enter into deliberation over what to believe. But even this seems not to be a case in which there are *epistemic* goals pulling against one another. Rather, it seems to be a case in which epistemic considerations pull against considerations that are altogether *non-epistemic*—such as the likely practical consequences of holding certain beliefs. Furthermore, it's not entirely clear that this is even a case of deliberation over what to believe, properly so speaking. Such cases may be best understood as involving deliberation over what to *desire* to believe or what to *try and bring oneself* round to believing, which seems a very different sort of activity.

Considering a broader range of examples might help to further motivate the claim that justification and knowledge are normatively coincident—as I've stated, though, my primary aim here is not to try and motivate this. Beyond this chapter, this assumption will play no essential role, but teasing out its consequences offers a shortcut to certain views that I will ultimately try to defend in other, more laborious, ways. If we do take this assumption seriously then, as I've suggested, a number of issues in epistemology get cast in a somewhat unfamiliar light. In particular, as I shall argue, the assumption has surprising consequences for how we ought to ascribe knowledge and justification in lottery cases and concomitant consequences regarding the relationship between justification and evidential probability (Smith 2014). In section 1.2 I will further clarify and explore the idea that knowledge and justification are normatively coincident goals. The consequences of this will be explored in sections 1.3 and 1.4.

1.2. The Normative Coincidence Constraint

To claim that two goals are normatively coincident, as we've seen, is to claim that one cannot aim at one of the goals without automatically aiming at the other. This implies, in turn, that it is impossible to aim for a situation in which one of the goals is achieved but the other is not. While normatively coincident goals can come apart, there is no method by which one could *cleave* them apart. If we suppose, as most epistemologists do, that justified belief is necessary for knowledge, it follows at once that one cannot aim for a situation in which one forms a belief that qualifies as knowledge but not as justified—for the situation itself could never obtain. If justification and knowledge are normatively coincident, however, then the reverse project will also be impossible. If justification and knowledge are normatively coincident then it is not possible for one to aim for the (possible) situation in which one forms a belief that qualifies as justified but not as knowledge.[2] Call this the *normative coincidence constraint.*

A few points of clarification are in order. First, it is standard to distinguish *epistemic* justification from justification that is *practical* or *prudential* (see, for instance, Jenkins 2007). Suppose once again that I'm about to take a driving test. Suppose that, in spite of failing so many times in the past, I have managed to convince myself that this time will be different and I'll finally pass. There is a sense in which holding this belief may well be in my overall best interests—it will enable me to approach the test more calmly and with more confidence and might increase the chances of my passing. For this reason, the belief might be described as prudentially justified. Nevertheless, the belief is clearly not an epistemically justified one—it is a poor fit with my evidence.

It is of course epistemic, and not prudential, justification that is supposed to feature in the normative coincidence constraint. For what it's worth, it seems clear that prudential justification is *not* something that normatively coincides with knowledge. As the driving test example shows, it is possible to form beliefs in a way that is aimed at prudential justification and not at knowledge. If we accept that epistemic justification is normatively coincident with knowledge, then this could perhaps provide a principled way of distinguishing between justification of the two kinds.[3]

[2] Some epistemologists have defended the view that a belief can qualify as knowledge without being justified—one example is Audi (1995, 2001). This is still compatible, in principle, with the claim that knowledge and justification are normatively coincident goals—that one cannot *aim* for one of these without aiming for the other also. Audi's view, though, would clearly seem to open up the possibility of aiming for knowledge without aiming for justification.

[3] While the distinction between epistemic and prudential justification is widely accepted, there is substantial disagreement as to how it should best be analysed. Some have proposed that what sets epistemic justification apart from prudential justification is the fact that it is conditioned by evidence (Williamson 2000: s. 9.8, Conee and Feldman 2004). Others have claimed that what makes justification epistemic is the fact that it is 'truth-linked' in a way that prudential justification is not (Alston 1985, Lemos 2007: 13–14). Still others have suggested that epistemic justification serves to promote a specialised set of 'epistemic' interests, while prudential justification serves to promote one's overall interests (Hazlett 2006). Characterising epistemic justification in terms of its normative coincidence with knowledge should, I think, be regarded as a further viable alternative. Another possibility is

Even epistemic justification, though, is standardly thought to come in a number of varieties. To put my cards on the table, the kind of epistemic justification that I take to enter into the normative coincidence constraint is *all things considered, doxastic* justification. Doxastic justification is, quite simply, justification conceived as an attribute of a subject's beliefs. It is usually contrasted with *propositional* justification—which is understood as a kind of relation between a subject and a proposition (see for instance Firth 1978: 217–18, Kvanvig and Menzel 1990). Often when we engage in epistemic evaluation we are interested in assessing the beliefs that people actually hold—we may say things like, 'The detective justifiably believes that the butler committed the murder' or 'The detective is justified in believing that the butler committed the murder.' Sometimes, though, we may want to evaluate a person's epistemic position with respect to a proposition, in abstraction from whether or not he actually believes it. Sometimes we say things like, 'The detective *has justification for believing* that the butler committed the murder.' We might say this if the detective has some good evidence incriminating the butler, even if he hasn't yet made up his mind or has jumped to the wrong conclusion.

The connection between doxastic and propositional justification is usually taken to be mediated by the *basing* relation. According to a now rather standard picture, in order for one to justifiably believe a proposition P one must, first, have justification for believing P and second, believe P in a way that is appropriately based upon that justification (see for instance Pollock 1986: 81, Korcz 1997). Returning to our detective, in order for him to justifiably believe that the butler committed the murder he must believe it, and believe it on the basis of the incriminating evidence he possesses.

What, then, is *all things considered* doxastic justification? To stick with the present example, we might imagine that the detective possesses some evidence that incriminates the butler while, at the same time, possessing other evidence that appears to exonerate him. Suppose the butler has a motive and that some of his fingerprints were found at the crime scene. But suppose it later emerges that the butler also has a solid alibi. In this case, we might say that the detective has *prima facie* justification for

to put forward the normative coincidence of justification and knowledge as a way of (partially) *interpreting* some of these other turns of phrase—as a way of interpreting the claims that justification is conditioned by evidence, or truth linked or that it promotes one's epistemic interests.

believing that the butler committed the murder—after all, he does possess evidence that weighs in support of this proposition. We might also say, though, that the detective does not have 'ultima facie' or *all things considered* justification for believing that the butler committed the murder—once we take all of his evidence into account, the proposition is not supported.

One has prima facie justification for believing P iff one possesses evidence that adequately supports P. One has all things considered justification for believing P iff, in addition, P is adequately supported by one's total body of evidence (see for instance Pollock 1970: 72–8, Alston 1985: 104–5, Señor 1996). If one believes P then the belief will count as prima facie justified iff it is appropriately based upon a prima facie justification for believing P. The belief will count as all things considered justified if, in addition, one's prima facie justification for believing P is not outweighed, or otherwise defeated, by any further evidence that one possesses.

It is important to stress that the kind of justification involved in the normative coincidence constraint is all things considered and not merely prima facie. If I were to believe a proposition on the basis of evidence that is defeated by further evidence I possess, then this could be a way of aiming for prima facie justification without aiming for knowledge—this would, in effect, be a cleaving method with respect to the two goals. But this would not, of course, be a way of aiming for all things considered justification and, thus, could not serve to cleave knowledge from *it*. The very fact that one could achieve prima facie justification in the way described strongly suggests that it shouldn't be regarded as a worthy epistemic goal in and of itself. Its role, rather, is that of a necessary condition for the attainment of worthy epistemic goals, such as all things considered justification and knowledge. Henceforth, I shall use 'justification' to mean all things considered justification, unless noted.

If the normative coincidence constraint is correct, then it is impossible for me to aim for a situation in which I justifiably believe a proposition without knowing it. For all that the constraint says, though, it may well be possible for me to aim for a situation in which *someone else* justifiably believes a proposition without knowing it. Indeed, this would seem to be possible—telling someone a lie might be a simple way of bringing such a situation about. I can treat others' beliefs as states in the world to be controlled for various ends. The normative coincidence constraint may go some way towards capturing the idea that I cannot treat my *own* beliefs in this way.

This observation does, however, point us towards certain, somewhat contrived, counterexamples to the normative coincidence constraint as stated—for there are certain, somewhat contrived, situations in which one's attitude towards one's future self can take on the character of an attitude to another. Suppose I know that I'm about to have my memory erased and I deliberately plant, for my future self, some evidence that misleadingly supports a proposition I know to be false. This would seem to be a viable way of aiming for a situation in which my future self, suitably estranged from my present self, forms a belief that qualifies as justified but not as knowledge. As I've suggested, these sorts of cases are better assimilated to the class of cases in which one attempts to manipulate the beliefs of another—in a way it is incidental that the target of the manipulation happens to be a disconnected future self. In any case, I shall set these cases to one side for the purposes of the present discussion.

1.3. Lottery Cases

In this section I will begin exploring some of the consequences of the normative coincidence constraint as I see them. One consequence concerns *lottery cases* like the following: suppose I hold a single ticket—ticket #542 say—in a fair 1,000-ticket lottery. Suppose the winning ticket has been drawn and, as it happens, it's ticket #457, but I'm yet to hear the result. In spite of this, suppose I believe, purely on the grounds that there are 999 losing tickets and only one winner, that my ticket—ticket #542—has lost. Both of the following claims are widely held amongst epistemologists:

(i) I do not know that ticket #542 has lost.
(ii) I am justified in believing that ticket #542 has lost.

Defenders of (i) include Harman (1968), Dretske (1971), DeRose (1996), Nelkin (2000), Williamson (2000: chap. 11), Hawthorne (2003: esp. chap. 1), and Pritchard (2005: 162–3, 2007) amongst others and defenders of (ii) include Kyburg (1970), Klein (1985), Foley (1993: chap. 4, 2009), Lewis (1996), Hawthorne (2003: 8, 9), and Pritchard (2007) amongst others. Both claims do, undoubtedly, have a certain intuitive appeal.

Many philosophers agree that, before hearing the lottery result, it would be inappropriate to assert that ticket #542 has lost, purely on the basis of the odds involved. This observation is often taken to support the *knowledge account* of assertion, according to which our practice of making assertions is governed by the rule that one should not assert a proposition unless one knows it (Williamson 2000: s. 11.2, Hawthorne 2003: s. 1.3). But this observation only favours the knowledge account on the assumption that (i) is true—and it only favours it over the rival *justified belief account* on the assumption that (ii) is true also. According to the justified belief account, our practice of making assertions is governed by the rule that one should not assert a proposition unless one justifiably believes it (see Kvanvig 2009, Smith 2012). Unless (ii) is true, the justified belief account will equally well predict that I cannot appropriately assert that ticket #542 has lost purely on the basis of the odds involved.

Claim (ii) does set the stage for the so-called 'lottery paradox'—at least when combined with two further assumptions. The first of these is a kind of symmetry principle according to which, if I have justification for believing that ticket #542 has lost the lottery then I must simultaneously have justification for believing that ticket #1 has lost, for believing that ticket #2 has lost, for believing that ticket #3 has lost, and so on for each of the 1,000 tickets. The second of these is the principle that justification is closed under multiple premise deductive consequence—the principle according to which, if one has justification for believing each of a set of premises, and these premises together deductively entail a conclusion, then one has justification for believing the conclusion.

If I have justification for believing that ticket #1 has lost, that ticket #2 has lost, that ticket #3 has lost . . . that ticket #1,000 has lost then, by multiple premise closure, I have justification for believing that all the tickets have lost. But I also have justification for believing that one of the tickets has won. By a further application of multiple premise closure, we derive the paradoxical conclusion that I have justification for believing an out and out contradiction; that all the tickets have lost and one of the tickets has won. Some epistemologists have rejected (ii) as a way of avoiding the lottery paradox. Others, though, see this reasoning as primarily casting doubt upon the other assumptions—particularly multiple premise closure.

How, then, does the normative coincidence constraint bear upon (i) and (ii)? In a way the answer is simple—if the normative coincidence constraint holds, then (i) and (ii) cannot *both* be correct. If (i) and

(ii) were both correct, then believing that ticket #542 has lost, on the basis of the odds involved, would be a way of aiming for justification but not for knowledge—it would be a 'cleaving' method with respect to the two goals. If knowledge and justification are normatively coincident then there can be no such method.

In drawing this conclusion, though, we do need to tread with some care. Cases in which a justified, true belief falls short of knowledge are very familiar and, presumably, their existence poses no general threat to the normative coincidence constraint. Why, then, should it make any difference if we choose to include *lottery* cases amongst their number? The answer, I think, is that there are significant differences between lottery cases and Gettier cases of a traditional sort. Consider, for contrast, the following standard Gettier case (originally described by Skyrms 1967): suppose I'm holding a 'Sure Fire' brand match, ready to strike it. I know that the match is dry and the environment is normal. Furthermore, I have used any number of Sure Fire matches in the past and they have always lit on the first attempt. I strike the match and form the belief that it's about to light. In actual fact, the match I'm holding, unlike a normal Sure Fire match, has a tip with a high proportion of impurities—so much so that it could never be lit just by friction. Nevertheless, the match *is* about to light, due to a coincidental burst of anomalous Q-radiation. In this case my belief that the match is about to light is justified and true and yet, intuitively, I don't know that the match is about to light.

In both the lottery case and the Gettier case, my belief fails to qualify as knowledge. In the Gettier case, however, this failing is clearly attributable to *extenuating circumstances*. What prevents my belief from qualifying as knowledge, in the Gettier case, is the abnormal condition of the match that I'm holding. Had the match been a normal one, my belief *would* have qualified as knowledge. Although knowledge and justification come apart in the Gettier case, they are driven apart by circumstances of which I have no awareness and over which I have no control. It is for this reason that the Gettier case supplies no cleaving method. My situation in the Gettier case is one that could *befall me*—but it's not one that I could aim for.[4]

[4] What if I were made aware of the impurities in the match but stubbornly continued to believe that the match was about to light? Would this be a way of aiming for the Gettier situation? It would not, since this new evidence would crucially alter the case—and, indeed,

In the lottery case, in contrast, there don't appear to be any extenuating circumstances that prevent my belief from qualifying as knowledge. What would such circumstances be? The lottery was drawn as expected, the proceedings were fair, everything was above board, my ticket did in fact lose etc. How exactly could the situation be made any *more* conducive for the belief that I form? But if the failure of my belief to qualify as knowledge is not attributable to extenuating circumstances of any kind then the buck, as it were, must stop with me and the way that I formed the belief. If we insist that my belief in the lottery case is justified, then we have before us the blueprint for a cleaving method. If knowledge and justification come apart in the lottery case, they are driven apart by *me* and not by circumstance. My situation in the lottery case *is* one that I could aim for.

The lottery case and the Gettier case, then, are significantly different. And, needless to say, there is nothing particularly special about the Sure Fire match case that I chose—any standard Gettier case could have been used to illustrate the point. All standard Gettier cases involve extenuating circumstances of one kind or another that lie beyond the ken and control of the protagonist. As such, the divergence of knowledge and justification in such cases is in no tension with the normative coincidence constraint. If the normative coincidence constraint is correct, though, then our attributions of knowledge and justification in the lottery case must converge—this is simply not the sort of case in which they could come apart. Either I cannot know that ticket #542 has lost, and cannot justifiably believe it either, or I can justifiably believe it, and can know it.

There is, I think, a general and pervasive picture of justification that lies behind the acceptance of (ii). The picture is something like this: for any proposition P we can always ask how *likely* it is that P is true, given present evidence. The more likely it is that P is true, the more justification one has for believing that it is. The less likely it is that P is true, the less justification one has for believing that it is. One has justification *simpliciter* for believing P when the likelihood of P is sufficiently high and the risk of ~P is correspondingly low. Securing justification for

alter it in such a way that it would no longer count as a Gettier case. In the new case, my belief that the match is about to light will still fail to qualify as knowledge—but it will also fail to qualify as justified. After all, I would be neglecting relevant defeating evidence. My belief would still qualify as justified prima facie but, as already discussed, the existence of a method that cleaves apart knowledge and prima facie justification should come as no surprise.

believing a proposition, on this picture, is essentially a matter of minimising one's risk of error—call it the *risk minimisation* conception of justification. This sort of view seems to be widely held amongst epistemologists. Indeed, it seems to be shared even by epistemologists who otherwise disagree quite profoundly about the nature of justification (see, for instance, Russell 1948: chap. VI, Chisholm 1957: 28, Derksen 1978, Alston 1988, Moser 1988, Plantinga 1993: chap. 9, Fumerton 1995: 18–19, Lewis 1996: 551, Swinburne 2001: chap. 3, 2011, Conee and Feldman 2004: n. 32, Pryor 2004: 350–1, 2005: 181, BonJour 2010, Goldman 2011: s. 16.7). In believing that ticket #542 has lost the lottery I am clearly running a very low risk of error. The evidence on which my belief is based is that there are 999 losing tickets and only one winner. The probability that ticket #542 has lost, given this evidence, is 0.999 and the probability that it has won is 0.001. If securing justification is really just a matter of minimising one's risk of error, then such a belief seems entirely beyond reproach.

There are a number of different ideas about knowledge that could potentially motivate the acceptance of (i). First, and most simply, there is the thought that knowledge requires the complete *elimination* of error risk. On this sort of view, in order to know a proposition P, P must be *certain* given one's evidence (Klein 1995, Williamson 2000: chap. 10). Clearly, my evidence in the lottery case doesn't make it certain that ticket #542 has lost. If we combine the claim that risk minimisation is sufficient for justification with the claim that risk elimination is necessary for knowledge, we derive an overall picture on which knowledge and justification are not normatively coincident goals. Lottery cases can be used to make this vivid.

One needn't think, though, that knowledge requires evidential certainty in order to motivate (i)—(i) could equally be motivated by the imposition of a *sensitivity* or *safety* requirement upon knowledge (Dretske 1971, Nozick 1981: chap. 3, Williamson 1992, 2000: chap. 5, Sosa 1999a, 1999b, Pritchard 2005: chap. 6, Smith 2009). Suppose one truly believes P based upon evidence E. E might be described as sensitive evidence for P iff had P been false then E would have been false. This, in turn, is sometimes cashed out in terms of the following possible worlds condition: in all of the closest or most similar possible worlds in which P is false, E is false. Further, E might be described as safe evidence for P iff P could not easily have been false, given that E is true. This, too, can be cashed out in terms of possible worlds: in all very close or similar possible worlds in which E is true, P is true. A belief might be described as sensitive or safe just in case it is based upon sensitive or safe evidence respectively. The sensitivity and safety conditions offer, I think, slightly different ways of capturing the requirement that a belief be attuned or responsive to the worldly condition that makes it true.

My belief that ticket #542 has lost the lottery, based upon the odds involved, will satisfy neither condition. Consider sensitivity first. If ticket #542 had won the lottery, then this would not have impinged upon my evidence or beliefs at all. By stipulation I have heard nothing about the lottery draw—my evidence for thinking that ticket #542 has lost is just that there are 999 losing tickets and only one winner, and this would still be the case if ticket #542 had won. In the most similar possible worlds in which ticket #542 wins the lottery, I still possess my evidence and still believe, now erroneously, that ticket #542 has lost.

My belief, then, is not sensitive. Neither is it safe. Ticket #457 may have actually won the lottery, but it could have easily lost—the lottery was not rigged and the ticket was not ordained to win etc. Furthermore, ticket #542 could have just as easily won as any other ticket—there were no special impediments or obstacles to ticket #542 winning. The circumstance in which ticket #542 won, then, could have easily come about and, as we've seen, this would have been a circumstance in which my evidence still holds. As long as there are similar possible worlds in which ticket #457 loses, there must be similar possible worlds in which other tickets win. But in this case, given that all the tickets are on a par, there must be similar possible worlds in which each of the tickets wins—including ticket #542. In these worlds, I still possess my evidence and still believe, now erroneously, that ticket #542 has lost.[5] My belief that ticket #542 has lost is neither sensitive nor safe.

If we combine the claim that risk minimisation is sufficient for justification with the claim that either sensitivity or safety is necessary for knowledge, we derive an overall picture on which knowledge and

[5] Some epistemologists have suggested a watered down safety condition that requires only that P be true in *most* of the close worlds in which E holds. This condition could presumably be met by my lottery belief in which case this alone will not be enough to motivate (i). In *Epistemic Luck* Duncan Pritchard describes a principle like this (see Pritchard 2005: 156) before opting for a strengthened safety condition that he takes to preclude lottery beliefs from qualifying as knowledge.

justification are not normatively coincident. Once again, lottery cases can be used to make this vivid. Whether or not the normative coincidence constraint makes any predictions about the nature of justification or the nature of knowledge, it does make predictions about how the two natures must be *coordinated* with one another. With the normative coincidence constraint in place, certain widespread views of justification and of knowledge cannot be combined.

1.4. A Closer Look at Lottery Cases

When it comes to attributing knowledge and justification in the lottery case, there are four initial options: either I know that ticket #542 has lost and I justifiably believe it, I don't know that ticket #542 has lost and I don't justifiably believe it, I don't know that ticket #542 has lost and I do justifiably believe it, or I know that ticket #542 has lost and I don't justifiably believe it. The final option can be ruled out on the grounds that justification is necessary for knowledge. As argued in section 1.3, the normative coincidence constraint allows us to rule out the penultimate option.

It's tempting to think that this is where the normative coincidence constraint will leave things. If knowledge and justification are normatively coincident then, as discussed, the lottery case is not the kind of case in which they could diverge. But all that follows from this is that we must be even-handed in the way that we attribute knowledge and justification, and there are two ways of doing this; we must either be maximally *generous*—granting that the belief that ticket #542 has lost is both justified and constitutes knowledge (Hill and Schechter 2007, Lycan 2006: n. 23, Reed 2010, see also Weatherson 2014: s. 4.4.2)—or maximally *stingy*—insisting that the belief that ticket #542 has lost falls short of knowledge and of justification too (Ryan 1996, Nelkin 2000, Smith 2010a, Smithies 2012).[6] Presumably, the normative coincidence constraint won't allow us to adjudicate between these two remaining options. It would be surprising

[6] Sutton, too, defends this stingy stance. Indeed, one of Sutton's motivations for the claim that justified belief *is* knowledge is that it motivates the verdict that one cannot justifiably believe, on the basis of the odds involved, that a particular ticket has lost a fair lottery and, thus, offers a way out of the lottery paradox (see Sutton 2007: s. 2.1.2). As I will show, this verdict can be ensured by far less dramatic claims (a point that much of the recent literature on the lottery paradox in any case serves to underline).

if this constraint alone allowed us to draw conclusions about the justificatory status of beliefs about lottery outcomes (or any beliefs for that matter).

While there is a sense in which this impression is correct, there is also a sense in which it turns out to be wrong. The normative coincidence constraint, as formulated in section 1.2, is crucially ambiguous—it permits of a weaker and a stronger construal. While the weaker construal is indeed neutral between the generous and stingy stances, the stronger construal is not—it weighs against the generous stance, leaving the stingy stance as the last option standing.

In section 1.2, the normative coincidence constraint was stated like this: it is not possible for one to aim for a situation in which one forms a belief that qualifies as justified, but not as knowledge. The ambiguity here is simply one of scope—more precisely, it concerns the relative scope of the modal operator and the existential quantifier. Let A be read 'It is possible for one to aim for a situation in which . . . is true', let J be read 'One justifiably believes . . . ', let K be read 'One knows . . . ', and let X be a variable ranging over propositions. The normative coincidence constraint could, I suggest, be formalised in either of the following two ways:

$$\text{(NCC1)} \quad {\sim}\exists X\ A(JX \wedge {\sim}KX)$$

$$\text{(NCC2)} \quad {\sim}A(\exists X\ (JX \wedge {\sim}KX))$$

According to NCC1, there is no proposition such that one can aim for a situation in which one justifiably believes it but doesn't know it. According to NCC2, one cannot aim for a situation in which there is some proposition that one justifiably believes but doesn't know. We might say that NCC1 forbids us from aiming *de re* to justifiably believe a proposition without knowing it, while NCC2 forbids us from aiming for this outcome *de dicto*.

NCC1 is the weaker of the two claims. If there is some proposition that I can aim to justifiably believe without knowing, then I can clearly aim for a situation in which I justifiably believe some proposition without knowing it. But the converse does not hold. If I can aim for a situation in which I justifiably believe some proposition without knowing it, it does not follow that there is some proposition that I can aim to justifiably believe without knowing. Accepting NCC1 and denying NCC2 seems a stable position—but it is difficult to see what might compel one to settle upon it. The simple thought that deliberation over what to believe does not involve the evaluation of competing epistemic goals would appear, for instance, to take us all the way to NCC2.

If NCC2 fails, then there must be a method for justifiably believing some proposition without knowing it. Even if this proposition cannot be specified in advance, such a method clearly secures justification at the expense of knowledge. Deliberation over whether to implement such a method could lead to deliberation over the relative value of the two goals. And this, or so it seems to me, is a place where deliberation over what to believe never leads.[7]

NCC1, however, is enough to support the consequences that have been drawn from the normative coincidence constraint so far. In particular, NCC1 will be enough to block the combination of claims (i) and (ii). If (i) and (ii) hold then there will be a proposition—namely, the proposition that ticket #542 has lost—that I can aim to justifiably believe without knowing. More formally, if L is the proposition that ticket #542 has lost, we have it that $A(JL \wedge {\sim}KL)$—which is inconsistent with NCC1. But this is as far as NCC1 will take us—it holds no further lessons as to how the lottery case should be described. As suggested above, though, there is more yet that might be extracted from NCC2.

Suppose I am justified in believing that ticket #542 has lost the lottery, based purely on the odds involved. Given that every ticket has an equally slim chance of winning, I would presumably be justified in believing the same thing about each of them. This is, in effect, the *symmetry principle* used in the above formulation of the lottery paradox.[8] But, however

[7] There may be certain cases that discriminate between NCC1 and NCC2. Suppose I ask a friend to arrange for me to encounter evidence that misleadingly favours some falsehood. When I do this, there's no specified proposition that I'm aiming to justifiably believe without knowing—thus, the case poses no threat to NCC1. But am I aiming to justifiably believe some unspecified proposition without knowing it, in defiance of NCC2? Perhaps—but matters are not clear-cut. If I really do have good reason to believe that misleading evidence has been planted for me, then I cannot justifiably continue to form beliefs as I usually would. In particular, if I encounter evidence that my friend could have conceivably planted then, given my background evidence, it seems that I wouldn't be justified in trusting the evidence without taking steps to rule this possibility out. If I did trust the evidence and simply believed whatever proposition it appeared to favour, then the belief may well fall short of knowledge, but it would seem also to be unjustified. I admit, however, that the treatment of cases like this is not entirely obvious and they could provide some motivation for accepting NCC1 without NCC2. I am indebted here to Aidan McGlynn.

[8] While this principle is widely accepted, some—such as Harman (1986: 70–2)—have taken the lottery paradox to cast doubt upon it (see also Douven 2008, Kroedel 2012). On the picture that Harman suggests, I can justifiably believe that ticket #542 has lost and that

generous we wish to be in appraising these 1,000 beliefs, it's clear that they could not *all* constitute knowledge. One of these beliefs, after all, has to be false. Given that knowledge is factive, the maximum number of these beliefs that could possibly constitute knowledge will be 999.

By forming these 1,000 beliefs I would, in effect, be guaranteed to form some belief that is justified but falls short of knowledge—naturally, though, I could not specify in advance which belief this would be. While this does not describe a method of justifiably believing any *particular* proposition without knowing it, it does describe a method for justifiably believing *some proposition or other* without knowing it. More formally, if L_n is the proposition that ticket #n has lost, we have it that $A((JL_1 \wedge \ldots \wedge JL_{1000}) \wedge ({\sim}KL_1 \vee \ldots \vee {\sim}KL_{1000}))$. But if it is possible for one to aim for a situation in which $(JL_1 \wedge \ldots \wedge JL_{1000}) \wedge ({\sim}KL_1 \vee \ldots \vee {\sim}KL_{1000})$ then it is possible for one to aim for a situation in which $\exists X\ (JX \wedge {\sim}KX)$, giving us $A(\exists X\ (JX \wedge {\sim}KX))$, which is inconsistent with NCC2, though consistent with NCC1. If I can aim for a situation in which each of a range of propositions is justifiably believed, but some amongst their number are not known, then I can aim for a situation in which I justifiably believe some proposition without knowing it. If NCC2 is to be maintained, our original assumption must be rejected—I am not justified in believing, on the basis of the odds involved, that ticket #542 has lost.

In order to elicit the result that I lack justification for believing that ticket #542 has lost, NCC1 must be combined with certain substantial claims about knowledge—such as the claims that knowledge requires safety or sensitivity or evidential certainty. NCC2, however, seems capable of generating this conclusion almost by itself. The only assumption

ticket #712 has lost etc. but I cannot believe this about all of the tickets *at once*. On this view, there will be a certain *number* of these beliefs that I can justifiably form, but any that I form in excess of this quota will not be justified. If I form too many of these beliefs then, even though each is equally supported by my evidence, some will be justified and some not, owing to the *order* in which they were formed. This picture, I suspect, is also in conflict with NCC2—but I won't argue this here.

If one held, with Sutton, that justified belief is knowledge, or, more generally, that justified beliefs must be true (Littlejohn 2012, Steglich-Petersen 2013), then one could adopt the following view: I can justifiably believe, of every ticket, that it has lost, with the exception of ticket #457 (being the ticket that has actually won). This would be another kind of view that violates the symmetry principle (thanks here to Kenny Easwaran). As far as I'm aware, such a view has never been defended. Sutton himself, as noted in n. 6, opts to deny that one can justifiably believe, of any ticket, that it has lost a fair lottery purely on the basis of the odds involved.

about knowledge that is needed is an assumption of factivity—that one can only know truths. And the only other assumption used is the symmetry assumption flagged above—if one can justifiably believe, of a given ticket, that it will lose the lottery on the basis of the odds involved, then one can justifiably believe the same about each ticket.

As discussed in section 1.3, the claim that I have justification for believing that ticket #542 has lost is underwritten by a certain widespread conception of justification—namely, the risk minimisation conception. As such, while NCC1 weighs against combining the risk minimisation conception of justification with certain requirements on knowledge—such as safety, sensitivity, or evidential certainty—NCC2, surprisingly enough, weighs against the risk minimisation conception all by itself.[9] As the lottery case illustrates, the interests of risk minimisation can, if the setting is right, be well served by believing each of a set of propositions that is known to be inconsistent. In general, if my evidence suggests that some member of a set of propositions is false but a high proportion of its members are true, then this is quite compatible with each member having a high evidential probability. Indeed, if this is my only relevant evidence about the propositions then, as the proportion of truths tends towards 1, so too will the individual evidential probabilities. If my epistemic goal is to ensure that my beliefs, taken individually, are very likely to be true, then I could scarcely do better than to believe every member of the set.

Clearly, though, by believing every member of such a set I effectively ensure that some of the beliefs I form will be false and will thus fall short of knowledge. By adopting this policy, I sacrifice one or more beliefs, condemning them to falsehood, for the sake of the many that will turn out to be true. Knowledge and high evidential probability are not normatively coincident—at least not in the sense captured by NCC2. If high evidential probability is sufficient for justification, as the risk minimisation conception would have it, then justification and knowledge are not normatively coincident—it will be possible to aim *de dicto* for a situation in which

[9] The only other assumption needed is the assumption that knowledge is factive. In this dialectical setting, the symmetry principle need not be regarded as an additional component—since it is, in effect, guaranteed by the risk minimisation conception. If I have justification for believing that ticket #542 has lost then, since it's just as likely on my evidence that ticket #1 has lost, ticket #37 has lost etc., the risk minimisation conception predicts that I must also have justification for believing each of these things.

I justifiably believe a proposition without knowing it. If NCC2 is correct then, to put it somewhat figuratively, every belief has a 'right' to aspire to the status of knowledge and, by adopting a policy that inevitably results in beliefs that could never achieve this status, I fail to respect that right.

It's worth noting, however, that NCC2 need not force us to give up on the risk minimisation conception *altogether*. While NCC2 is in conflict with the claim that high evidential probability is *sufficient* for justification, it is compatible with the claim that high evidential probability is *necessary* for justification. It could well be that NCC2 allows us to salvage something recognisably similar to the original risk minimisation picture—a kind of refinement of the picture rather than something completely different.

It will turn out, though, that no condition that is defined exclusively in terms of evidential probabilities, and demands less than evidential certainty, could suffice for justification compatibly with NCC2—some kind of new ingredient will need to be introduced. If ϕ is a property that is defined exclusively in terms of evidential probability and is satisfied by some propositions that are less than evidentially certain then, given relatively innocuous underlying assumptions, it can be shown that ϕ can be simultaneously satisfied by each member of an inconsistent set of propositions. This is a corollary of a result to be proved in Chapter 9. It follows that, if ϕ is a property that is defined exclusively in terms of evidential probability, and is satisfied by some propositions that are less than evidentially certain, it will be possible to aim *de dicto* to believe, but not know, a proposition that satisfies ϕ.

If one resolved to stand by the risk minimisation conception of justification, then one could of course treat all of this as reason to go back and reject the normative coincidence constraint. One could insist that it *is* after all possible to aim for justification without aiming for knowledge, and that lottery cases present an opportunity to do just this. While I think this would be the wrong reaction, I have perhaps said little to discourage it. One aim of this book is indeed to argue against the risk minimisation conception—but I don't take myself to have given much of an argument yet. My aim in this chapter has been more modest: to offer a glimpse of how some parts of epistemology might end up looking if we took the normative coincidence of justification and knowledge seriously as a starting point. It is surprising how much can be wrung from this seemingly innocuous principle. And it is telling that it, alone, seems enough to force justification and evidential probability apart.

2

What Justifies Belief

2.1. Risk Minimisation

There are two simple theses behind what I've been calling the *risk minimisation* conception of justification. One of these concerns categorical attributions of justification and the other concerns comparisons:

(a) One has justification for believing a proposition P iff P is likely, given one's evidence, to be true.
(b) One has more justification for believing a proposition P than a proposition Q iff P is more likely, given one's evidence, than Q is.

A further claim is needed in order to make these intelligible: a body of evidence furnishes all, or most, propositions with evidential probabilities—probabilities that are usually taken to be describable by a classical (Kolmogorovian) probability function.[1] This is, perhaps, more usefully

[1] A probability function is defined over a set of propositions that includes a 'maximal' proposition, entailed by each of the others in the set, and is closed under negation and disjunction. Propositions are standardly modelled, for this purpose, as subsets of a set of possible worlds W, with W itself playing the role of the maximal proposition. A classical probability function Pr is any function that assigns real numbers to the propositions in this set and satisfies Kolmogorov's three axioms (Kolmogorov 1933):

(i) $Pr(P) \geq 0$
(ii) $Pr(W) = 1$
(iii) $Pr(P \vee Q) = Pr(P) + Pr(Q)$, for any P and Q such that P and Q are inconsistent.

If the domain of the function is taken to be infinite, and closed under countable disjunction, then (iii) might be strengthened to the following:

(iv) If P_i, $i = 1, 2 \ldots$ is a sequence of pairwise inconsistent propositions then $Pr(\vee_i P_i) = Pr(P_1) + Pr(P_2) + Pr(P_3) \ldots$

A probability function that satisfies (iv) is known as *countably additive*.
It isn't an essential part of the risk minimisation conception that evidential probabilities be viewed as classical—though this would be the orthodox stance. There are many alternative approaches; views that allow for infinitesimal probabilities or for interval-valued probabilities, axiomatisations that treat conditional probabilities as primitive, axiomatisations that

thought of as a precondition for the intelligibility of theses (a) and (b) rather than a further thesis alongside them. Talk of a proposition being 'likely' in thesis (a) should be further clarified in terms of a proposition having an evidential probability that exceeds some threshold t that lies close to 1 and may be variable and/or vague[2]—but the details of this will matter little for the moment.

Some epistemologists have dissented from this picture and defended alternative views—the theories proposed by Sutton (2007), Bird (2007), Reynolds (2013), and Ichikawa (2014) have already been mentioned.[3] L. J. Cohen (1977: chap. 22, 1979), Dana Nelkin (2000), and Jarrett Leplin (2009) might also be added to this list. But, as noted in the last chapter, the majority of epistemologists do appear to accept that something along the lines of the risk minimisation conception is correct. And there is, undeniably, something very natural about it. After all, if justification does not require evidential certainty then what could it possibly require if not evidential likelihood? If justification does not demand the complete elimination of error risk then what could it possibly demand if not its minimisation? One could, when thinking in this way, almost get the impression that (a) and (b) are not substantial claims at all but, rather, serve to *define* a basic fallibilist notion of justification—a notion that comes before any substantial epistemic theorising.

The reflections of Chapter 1 show us, at the very least, that (a) and (b) cannot really be trivialities. Whatever we make of the claim that knowledge and justification are normatively coincident goals, it is not

embed non-classical logics etc. The risk minimisation conception, as I understand it, is quite compatible with approaches such as these.

[2] Any evidential probability value that one chooses is bound to seem rather arbitrary as a threshold for justification. Making the threshold variable and/or vague is sometimes thought to help with this—though it's not entirely clear how this is supposed to work. On the contrary, introducing a number of seemingly arbitrary values that the threshold can assume might be thought to compound the problem. Whatever the truth of the matter, I won't make much of this 'threshold problem' here—but I will say a little more about it in Chapter 5.

[3] Even Sutton, who holds that justification, strictly speaking, is identical with knowledge, thinks that something like the risk minimisation conception often captures what we intend to *convey* with justification attributions. According to Sutton, when we say that one justifiably believes P often we are just speaking loosely and mean only that one justifiably believes that P is *probable* (Sutton 2007: s. 2.3), which may be true if the evidential probability of P, for one, is sufficiently high. I am inclined to think that these two sorts of judgement are not nearly so close as Sutton supposes.

something that is *trivially* false. The idea that beliefs can be appraised as permissible or impermissible, in a distinctively epistemic way, is one that is betrayed by a broad range of ordinary practices; by our evaluating the adequacy of methods of inquiry and in determining when inquiry into a matter might reasonably cease, by our condemning assertions as ungrounded or premature and in our criticising actions for the beliefs upon which they are based. If there is a 'basic' notion of justification, it is captured by a platitude like this: one has justification for believing a proposition P iff one is permitted, epistemically, to believe P. Even if it is true that the minimisation of error risk is how we earn epistemic permission to believe things, it is not true by stipulation.

I'm inclined to think, however, that this is not true at all. In this chapter I shall outline an argument against the risk minimisation conception. I have already offered an argument of a kind against this conception—namely, one that proceeds from the assumption that knowledge and justification are normatively coincident goals. The argument I offer here will not use this as a premise—though it will appeal to it as a kind of supplementary consideration. As well as offering an argument against the risk minimisation conception, I shall begin the task of sketching a possible alternative picture. First, though, a little more stage setting is required.

I assume here that evidence is propositional—that one's body of evidence consists of a stock of propositions or a conjunction of propositions (Williamson 2000: s. 9.5, Dougherty 2011). I won't be defending any particular account of when a proposition qualifies as part of one's body of evidence—indeed, everything that I say here will be compatible with a number of different ways of thinking about this. In particular, everything that I say here can be reconciled with Williamson's knowledge account of evidence, according to which one's evidence consists of the propositions that one knows (Williamson 2000: chap. 9), and with more restrictive accounts that limit one's evidence to a proper subset of one's knowledge, such as the knowledge of one's own experiences and mental states (Lewis 1996, Conee and Feldman 2008, Swinburne 2011).[4]

[4] Some epistemologists hold that one's evidence fundamentally consists of one's experiences and mental states *themselves*, rather than propositions about them. I have no particular objection to this way of speaking—but what I do think is that many of the relations that evidence bears to propositions have to be analysed, first and foremost, as

Even if we fix upon a particular account of evidence, there is still room for substantial disagreement over the nature of evidential probability—over what it takes for a piece of evidence to confer a particular probability value upon a proposition. *Externalists* about evidential probability hold that evidential probability values are, in general, the product of contingent facts—in particular, facts about *propensities* or *frequencies* (see Russell 1948: chap. VI, Alston 1988: s. I, Plantinga 1993: chap. 9, s. I). On one sort of externalist view, the probability of a proposition P, given evidence E, will be determined by the frequency with which the kind of circumstance described by E is accompanied by the kind of circumstance described by P across actual and similar counterfactual circumstances. *Internalists*, on the other hand, conceive of evidential probability values as reflecting necessary, internal connections between evidence and hypotheses (see Keynes 1921, Carnap 1950, Kyburg 1971, Moser 1988, Fumerton 1995, 2004, Williamson 2000: chap. 10, Conee and Feldman 2008: s. 1.5). Once again, I take no stand on this issue here—but I will have more to say about it in future chapters.

I am taking it for granted here that believing a proposition involves a commitment over and above merely regarding it as likely. In one sense this seems obvious, and there is little temptation to think otherwise—to claim that P is *likely* is precisely to avoid committing to its truth. I can regard it as *extremely* likely that, say, the number of stars in the universe is a composite number or that ticket #542 has lost the lottery without actually believing either of these things.[5] In his *Rules for the Direction of the Mind* (1628), Descartes famously advised that we should never believe that which is

inter-propositional relations. A proposition cannot literally be entailed or made probable, for example, by an experience *itself*, though it can be entailed or made probable by the *occurrence* of an experience—that is, by the proposition that the experience occurred (for discussion of this see Williamson 2000: s. 9.5, Dougherty 2011). Propositions about one's experiences and mental states must be regarded as evidence in at least a derivative sense, even if we want to hold on to a stricter sense of the term reserved for the experiences and mental states themselves.

[5] There are, of course, difficult questions about the relationship between outright belief and degrees of belief—and these are not questions that I mean to prejudge here. According to one view, sometimes dubbed the 'Lockean Thesis', outright belief corresponds to a high degree of belief—one counts as believing a proposition just when the degree of belief that one invests in it is suitably high (see, for instance, Foley 1992, 1993: chap. 4, 2009). I don't mean to assume here that such a view is incorrect—only that it cannot be combined with a view on which degrees of belief are equated with overt probability estimates. I shall have more to say about degrees of belief in section 8.4.

merely probable (rule II). I take it for granted that such advice is, at the very least, *intelligible* (Kaplan 1995: 121). (In fact, I am inclined to think that, when appropriately construed, Descartes's advice is perfectly *sound*—more on this later). If the evidential probability of proposition P given evidence E is high, then it is, in some sense, appropriate for any subject in possession of this evidence to offer a generous estimate of P's probability. Whether it's permissible for a subject, so situated, to believe that P is true is a further question—and one that may, as I hope to show, have different answers depending upon the character of E.

2.2. Problems for Risk Minimisation

One possible argument against the risk minimisation conception proceeds from the principle of *multiple premise closure*, mentioned in Chapter 1. It's very plausible that deduction, of all things, should be an epistemically permissible way of expanding one's set of beliefs—that deduction, of all things, won't take me from beliefs that I am epistemically permitted to hold to beliefs that are epistemically impermissible. One way of attempting to make this rough intuition precise is the following: if one has justification for believing each of a set of propositions, then one has justification for believing any proposition that deductively follows from them.

Multiple premise closure is in clear tension with thesis (a) of the risk minimisation conception. Multiple premise deductions can aggregate error risk—the risk of error to which I expose myself in believing the conclusion of a multiple premise deductive inference may be higher than the risk of error to which I expose myself in believing any of the premises, taken individually. That is, the conclusion of a multiple premise deduction can inherit the error risk associated with each of the premises and, thus, have an evidential probability that dips below the threshold for justification, even if the evidential probability of each premise exceeds it. The risk minimisation theory would appear to be compatible with the weaker principle of *single premise closure*: if one has justification for believing a proposition, then one has justification for believing any proposition that deductively follows from it. If a deductive inference has but a single premise then there is no error risk to aggregate.[6]

[6] The closure principle described here concerns *propositional* justification. We could also formulate a related closure principle for doxastic justification: if one justifiably believes

The fact that the risk minimisation conception predicts the failure of multiple premise closure does, I think, constitute one kind of objection to it. But the dialectical situation concerning multiple premise closure is complex, and I won't try to press this objection yet—though I will return to it in Chapter 4. The argument I will offer here involves a rather different prediction that the risk minimisation conception makes—a prediction concerning the force of evidence that is purely statistical in nature.

Consider the following example, adapted from one originally devised by Dana Nelkin (2000: 388–9): suppose that I have set up my laptop such that, whenever I turn it on, the colour of the background is determined by a random number generator. For one value out of one million possible values the background will be red. For the remaining 999,999 values, the background will be blue. One day I arrive at a desk in the library and turn on my laptop. Moments before the background appears, I spot my friend Bruce at a nearby desk and wander over to say hello.

Bruce is already working away on his laptop and, when I set eyes upon it, it simply *appears* to me to be displaying a blue background and I immediately come to believe that it is. Let's suppose, for the time being, that my relevant evidence consists of two propositions (I will consider other ways of describing the evidential situation in due course):

(E_1) *It visually appears to me that Bruce's laptop is displaying a blue background.*

(E_2) *It is 99.9999% likely that my laptop is displaying a blue background.*

Here are a few preliminary observations about this case: if I were to believe that my laptop is displaying a blue background before returning

each of a set of propositions, and deduces a further proposition from them, then this further proposition is justifiably believed. To avoid certain counterexamples, we should specify that the deduction is careful and competent and that the relevant beliefs remain justified while performing it etc. Even when suitably qualified, we would expect this principle to fail on the risk minimisation conception for the reasons given in the body text, though a single premise version might, once again, remain secure.

The connection between the propositional and doxastic closure principles is very close—though mediated by certain further assumptions. If one justifiably believes each of a set of propositions then one must have justification for believing each of these propositions and believe each in a way that is based upon its justification. If these propositions deductively entail a further proposition then the propositional closure principle guarantees that one has justification for believing it. All that is then needed for the doxastic principle to be assured is the assumption that deduction preserves proper basing—if one (competently) deduces a conclusion from premises that are properly based on their justifications (throughout the deduction), then that conclusion will be properly based upon its justification.

to my desk, it would be natural to describe this belief as a *presumption* (perhaps a very safe one), while it does not seem at all natural to describe my belief about Bruce's laptop in these terms. Second, my belief about Bruce's laptop would appear to be a very promising candidate for knowledge—indeed, it *will* be knowledge, provided we fill in the remaining details of the example in the most natural way. If I were to believe that my laptop is displaying a blue background, this belief would never constitute knowledge even if it happened to be true. If, for instance, my battery died before I got back to my desk, I might well think to myself, 'I guess I'll never know what colour the background really was.' But if Bruce's battery died I certainly wouldn't think this about the background colour on his laptop. To believe that Bruce's laptop is displaying a blue background would be to believe in a way that is straightforwardly directed at knowledge. To believe that my laptop is displaying a blue background would be to believe in a way that seems indifferent to knowledge.

If someone were to ask me, 'What colour is the background on Bruce's laptop?', I would be perfectly epistemically entitled to reply 'It's blue.' But if someone were to ask me the same question about *my* laptop, it seems as though I ought to be more circumspect, and say something along the lines of 'It's overwhelmingly likely that the background is blue—but I haven't actually seen it.' Presumably, this is what I ought to *believe* too. I'm not required to do any further investigation into the background colour displayed by Bruce's laptop—even though I easily *could* by, for instance, asking others to have a look. But I ought to do more investigation into the background colour displayed by my laptop—by, for instance, going and having a look myself—before I rest on my laurels.

The implication of these considerations seems clear enough: I have justification for believing that Bruce's laptop is displaying a blue background, but I don't have justification for believing that my laptop is displaying a blue background. In spite of this, the proposition that my laptop is displaying a blue background is more likely, given my evidence, than the proposition that Bruce's is. While E_1 does make it highly likely that Bruce's laptop is displaying a blue background, it clearly does not *guarantee* that it is. After all, I could be hallucinating, or I could have been struck by colour blindness, or I could be subject to some strange optical illusion etc. No doubt these are all rather unlikely scenarios—but presumably the likelihood, given evidence E_1, that Bruce's laptop is displaying a blue background would be nowhere near as high as

99.9999%. This, of course, is precisely how likely it is that my laptop is displaying a blue background, given evidence E_2.[7] In believing that Bruce's laptop is displaying a blue background, I am actually running a *higher* risk of error than I would be in believing the same thing about my laptop. If this set-up were replicated again and again then, in the long run, we would expect the belief about Bruce's laptop to be false more often than the belief about my laptop.

The judgement that I lack justification for believing that my laptop is displaying a blue background is in considerable tension with the risk minimisation conception. This proposition is clearly *very likely*, given my evidence—by believing this proposition I would be running only a minute risk of error. If the risk minimisation theory is correct, then I could scarcely do better, epistemically, than to believe a proposition like this.

Provided the likelihood threshold t is set below 0.999999, the judgement that I lack justification for believing that my laptop is displaying a blue background will conflict with thesis (a) of the risk minimisation conception. One could, perhaps, try to preserve (a) by arguing that the likelihood threshold t should, for whatever reason, be set very high in the case that I've described. Bringing into play the judgement that I *have* justification for believing that Bruce's laptop is displaying a blue background effectively blocks this kind of manoeuvre. Try as we might, we will never find a likelihood threshold such that the proposition that Bruce's background is blue lies on the right side of it and the proposition that my background is blue lies on the wrong side.

To refute thesis (b) of the risk minimisation, we needn't even make any categorical judgements about what I have justification for believing—it's enough that we make the *comparative* judgement that I have more or better

[7] One could think of this as an instance of David Lewis's 'principal principle' (Lewis 1980). Suppose my only evidence relevant to a proposition P is that the objective probability of P at time n = x. Suppose, in particular, that I have no relevant evidence pertaining to things that happened after n. According to the principal principle, roughly speaking, the evidential probability of P, for me, will also be equal to x. Time n, in this case, could be thought of as the time at which I turn on my laptop.

Whether my evidence, in the case described, should be interpreted, ultimately, as evidence pertaining to an *objective* probability value is debatable. I needn't take any stand on this—all that is important, for present purposes, is that its probabilistic bearing be clear-cut. It may be that it is best interpreted, at the end of the day, as evidence about a propensity or an expected frequency or some such. It's worth noting though that if we *ever* have evidence regarding objective probability values, then cases like the one described must be amongst the clearest cases in which we do.

justification for believing that Bruce's background is blue than I do for believing that my background is blue. Strictly, this is all that is required to refute the letter of the risk minimisation conception.

The laptop example prompts a cluster of judgements that don't seem to fit in well with the risk minimisation picture. It may be rather tempting, however, for one to simply disregard such judgements as confused or naive. Perhaps we are simply accustomed to relying upon perception when it comes to such matters and suspending any scruples about its fallibility. Once we do reflect on the various ways that perception can go wrong, so this thought goes, these troublesome intuitions are exposed as a kind of prejudice, and my justificatory standing with respect to the two propositions in question no longer seems so different (see Steglich-Petersen 2013: s. 3.3). I'm not entirely convinced that this is the wrong thing to say—but I do suspect that it is.

Consider another example—one that is well known in legal theory and the philosophy of law, though less so in epistemology: suppose a bus causes some harm on a city street—it damages a car or injures a pedestrian or some such. In the first scenario, an eyewitness to the incident testifies that the bus was owned by the Blue-Bus company. In the second scenario, there is no eyewitness, but there is some unusually strong statistical evidence regarding the distribution of buses in the relevant area—evidence to the effect that 95% of the buses operating in the area, on the day in question, were owned by the Blue-Bus company.

Testimony, as we all know, is not perfectly reliable—particularly when it comes to testimony concerning an event of this kind.[8] The eyewitness in scenario one could have suffered a hallucination or she could have fabricated her memory of the incident or she could have deliberately concocted a lie in order to smear the Blue-Bus company, etc. These possibilities may not be likely, but if we were forced to come up with some estimate of the probability that the bus involved really was a Blue-Bus bus, given the witness testimony, it's doubtful that we would go quite as high as 95%—this would seem overly trusting. But this, of course, is precisely how probable the proposition is, given the strong statistical evidence available in scenario two.[9]

[8] See, for instance, Loftus (1996).

[9] If one thought that 95% really is a reasonable estimate, in scenario one, of how likely it is that the bus involved was a Blue-Bus bus, we could simply increase the proportion of

In spite of this, so long as we don't have any positive reason to think that the eyewitness in scenario one is mistaken or lying, it would be perfectly reasonable for us to take this testimony at face value and conclude that it was a Blue-Bus bus that caused the incident. Indeed, if this is our only relevant evidence, it would seem reasonable for us to assert and to act upon this conclusion. But what about scenario two? Should I really believe that the bus involved was a Blue-Bus bus on the grounds that 95% of the buses in the area on the day in question were Blue-Bus buses? Should I go around announcing that the bus involved was a Blue-Bus bus? Should I take steps against the company—boycott their buses, picket their offices, etc? If my *only evidence* is that 95% of the buses in the area on the day in question were Blue-Bus buses then to take such steps would surely be unjust (see Redmayne 2008, Enoch, Fisher, and Spectre 2012, Buchak 2014). On the risk minimisation conception, though, any unwillingness to believe or to act in this scenario is very puzzling. After all, the proposition that the bus involved was a Blue-Bus bus is made *very likely* by the statistical evidence I possess. By believing this proposition I would only be running a very small risk of error—I would have managed the risk of error in an almost exemplary way.

Questions about the force of purely statistical evidence are not, of course, merely academic—and it is not just theorists who have converged on these answers. Under prevailing legal practice, across a range of jurisdictions, eyewitness testimony of the sort available in scenario one could serve as a legitimate basis for a finding to the effect that the bus involved was a Blue-Bus bus—and for a finding of liability in a civil trial. In contrast, it would not be considered legitimate to base a finding of liability upon the kind of statistical evidence available in scenario two. Indeed, the general reluctance of courts, and individuals, to base verdicts of guilt or liability on evidence that is purely statistical in nature is well established.[10] The defender of the risk minimisation picture could, of

Blue-Bus buses in scenario two to restore the intended overall structure of the example. The probability value in scenario two is derived using what is sometimes called *statistical syllogism*: if I have evidence to the effect that a is a G and that the proportion of Gs that are Fs is x and have no further evidence relevant to whether a is an F, then the evidential probability, for me, that a is an F will be equal to x.

[10] The Blue-Bus example in fact traces back to a genuine civil case—*Smith v. Rapid Transit Inc* 317 Mass. 469, 58 N.E. 2d 754 (1945). The trial court, in this case, found that the imposition of liability could not rest upon the kind of purely statistical evidence supplied by

course, simply put this pattern down to prejudice as well—some kind of unreflective, knee-jerk preference for testimonial over statistical evidence. Once again, this may turn out to be the right thing to say—but I shall experiment, in the next section, with a possible way of giving these intuitions and conventions more credit.

2.3. Introducing Normic Support

Turning back to the laptop example, consider again the relationship between the proposition (P_1) that Bruce's laptop is displaying a blue background, the proposition (P_2) that my laptop is displaying a blue background, and my available evidence. Clearly, my evidence entails neither P_1 nor P_2. It would be perfectly possible for E_1 and E_2 to both be true while P_1 and P_2 are both false. Notice, though, that if E_1 is true and P_1 is false, then this would appear to be a circumstance crying out for *explanation* of some kind. If I visually perceive that Bruce's laptop is displaying a blue background when in fact it is *not*, then there has to be some explanation as to how such a state of affairs came about. Possible explanations have already been floated above—perhaps I'm hallucinating, or have been struck by colour blindness, or am subject to an optical illusion, etc. It can't be that I *just misperceive*—there has to be more to the story.

The circumstance in which E_1 and P_1 are both true, we might say, is *explanatorily privileged* over the circumstance in which E_1 is true and P_1 is false. E_2 and P_2, however, do not appear to stand in this relationship. Although it would be very unlikely for P_2 to be false while E_2 is true, this is not something that would require special explanation. All of the

the complainant—a verdict upheld, on appeal, by the Supreme Judicial Court of Massachusetts. These verdicts are very much in keeping with a broader pattern in which purely statistical evidence has been judged insufficient to meet legal burdens of proof. Other cases that are sometimes cited in this regard include *Virginia & S.W. Ry. Co. v. Hawk* 160 F 348, 352 (6th Cir., 1908), *Evans v. Ely* 13 F.2d 62, 64 (3rd Cir., 1926), *Commercial Standards Insurance Co. v. Gordon Transports Inc.* 154 F.2d 390, 396 (1946), *People v. Collins* 438 P.2d 33 40–1 (Cal. 1968), *State v. Carlson* 267 N.W. 170, 179 (Minn. 1978), *United States v. Shonubi* 103 F 3.d 1085 (2d Cir., 1997), *R v. Watters* Court of Appeal, Criminal Division (19 October 2000). For further relevant references, research, and discussion see Kaye (1982: s. I), Wells (1992), and Allensworth (2009: s. IIB). Many have concluded, on the basis of the case law, that standards of legal justification cannot be understood in purely probabilistic terms—a direct analogue, in effect, of the view that I shall defend about standards of epistemic justification.

random numbers that might be generated by my laptop are on an *explanatory par*. The occurrence of the one 'red' number does not require any more explanation than the occurrence of the 999,999 'blue' numbers. This, indeed, is part of what is involved in conceiving of a process as genuinely *random*.

If my belief that Bruce's laptop is displaying a blue background turned out to be false then, given the evidence upon which it is based, there would have to be some explanation for the error—either in terms of perceptual malfunction or disobliging features of the environment. If my belief that my laptop is displaying a blue background turned out to be false then, given the evidence upon which it is based, there need *not* be any available explanation for the error—the buck, as it were, may simply stop with me and the way that I chose to form my belief.

We can draw a similar contrast between the two kinds of evidence at play in the Blue-Bus example. In scenario one, if it turned out that the bus involved was not owned by the Blue-Bus company, in spite of the eyewitness testimony, then there would have to be some accompanying explanation—the eyewitness was hallucinating or lying or had a fabricated memory. Whatever the truth of the matter, it can't *just so happen* that the testimony was wrong—there has to be more to the story. In scenario two, though, it could *just so happen* that the bus involved was not a Blue-Bus bus in spite of the fact that 95% of the buses in the area were. While this would, in a sense, be surprising, given the proportions involved, it clearly wouldn't demand any further special explanation.

The idea that *normalcy* is purely a matter of statistical frequency or propensity is, undeniably, an attractive one. Adopting it, though, forces us to give up on another attractive idea—namely, that normal conditions require *less explanation* than abnormal conditions do. Sometimes when we use the term 'normal'—when we say things like 'Rain is normal at this time of year'—we might be making a straightforward claim about statistical frequency. Other times—when we say things like 'Tim would normally be home by six' or 'When I turn my key in the ignition, the car normally starts'—part of what we are trying to express, I believe, is that there would have to be some satisfactory *explanation* if Tim wasn't home by six or the car wasn't starting.

In this sense of 'normal' it could be true that Tim is normally home by six, even if this occurrence is not particularly frequent. What *is* required is that exceptions to this generalisation are always explicable *as* exceptions by the citation of independent, interfering factors—his car broke down, he had a late meeting, he had to detour around roadworks etc. If this condition is met, then the best way to explain Tim's arrival time each day is to assign his arrival by six a *privileged* or *default* status and to *contrast* other arrival times with this default (see Pietroski and Rey 1995).

This may well be possible even if the number of occasions on which Tim arrived home by six is exceeded by the number of occasions on which he arrived home later. Suppose Tim is significantly delayed, day after day—first by car trouble, then by a late meeting, then by roadworks, etc.—but, were it not for these interfering factors, he would always arrive home by six. There's a sense of 'normal' on which it remains true that Tim *normally* arrives home by six—we could imagine saying 'Tim would normally be here by six, but he's just had a bad run lately!'

Say that a body of evidence E *normically supports* a proposition P just in case the circumstance in which E is true and P is false requires more explanation than the circumstance in which E and P are both true (Smith 2010a). Given my evidence in the laptop example, the circumstance in which Bruce's laptop is not displaying a blue background would require more explanation than the circumstance in which it is. In contrast, the situation in which my laptop is not displaying a blue background, as unlikely as that might be, would not require more explanation than the circumstance in which it is. My evidence normically supports the proposition that Bruce's laptop is displaying a blue background, but does not normically support the proposition that my laptop is displaying a blue background.

Turning to the Blue-Bus example, if we have evidence to the effect that an eyewitness testified that the bus involved in the incident was a Blue-Bus bus, then the circumstance in which it wasn't a Blue-Bus bus requires more explanation than the circumstance in which it was. The testimonial evidence normically supports the proposition that the bus involved was a Blue-Bus bus. If, on the other hand, we have evidence to the effect that 95% of the buses operating in the area were Blue-Bus buses, then the circumstance in which the bus involved wasn't a Blue-Bus bus requires no more explanation than the circumstance in which it was. Given that 95% of the buses operating in the area were Blue-Bus buses, it would *frequently* be true that the bus involved in the incident was a Blue-Bus bus. Given that an eyewitness testified that the bus involved was a Blue-Bus bus, it would *normally* be true that the bus involved was a Blue-Bus bus.

The distinction between these different sorts of evidential support might fruitfully be compared to the distinction between statistical generalisations and ceteris paribus generalisations widely accepted in the philosophy of science (see, for instance, Millikan 1984: 5, 33–4, Pietroski and Rey 1995: 1.2). It might also be compared to the distinction, widely recognised in the philosophy of language, between generics that contain frequency adverbs—like 'As are frequently B', 'As are typically B'—and generics that are 'unmarked'—generics of the form 'As are B' (see, for instance, Leslie 2008). These comparisons may be particularly apt if we are inclined to understand evidential probability along externalist lines. As discussed above, on the externalist conception of evidential probability, to say that a piece of evidence probabilifies a given proposition is, quite literally, to make a kind of statistical generalisation. I'll have more to say about such comparisons in Chapter 6.

I have characterised an evidential support relation that demands more than probability, but less than certainty. And it is not difficult to appreciate, at least in a rough and ready way, why this relation might have some connection with epistemic justification. If one believes that a proposition P is true, based upon evidence that normically supports it, then, while one's belief is not assured to be true, this much is assured: if one's belief turns out to be false, then the error has to be *explicable* in terms of disobliging environmental conditions, deceit, cognitive or perceptual malfunction, etc. In short, the error must be attributable to *mitigating circumstances* of some kind and thus *excusable*, after a fashion. Errors that do not fall into this category are naturally regarded as errors for which one must bear *full responsibility*—errors for which there is no excuse. And, if error could not be excused, then belief cannot be permitted.

What I propose is that, in order for one to have justification for believing a proposition, it must be normically supported by one's evidence. When one classifies a belief as justified, one is committed to the claim that, if the belief fails to be true, then this failure will be independently explicable in terms of some identifiable interfering factor. To borrow a turn of phrase used by Pietroski and Rey (1995: 84), the notion of justification answers to the need to idealise in a complex world, not the need to describe a chancy one. The reason that I have justification for believing that Bruce's laptop is displaying a blue background, but lack justification for believing that my laptop is displaying a blue background, is that the former proposition is normically supported by my evidence while the latter proposition is not.

2.4. Normal Worlds

Suppose that possible worlds can be compared for their normalcy.[11] Normic support could, then, be modelled in terms of variably restricted quantification over worlds: a body of evidence E normically supports a proposition P just in case P is true in all the most normal worlds in which E is true. Alternatively, we might say that E normically supports P just in case there is a world at which E is true and P is true which is more normal than any world at which E is true and P is false. In fact, these two characterisations are not quite equivalent—the former assumes that there will be maximally normal worlds in which E is true, for any conceivable body of evidence E. I'll make this assumption for now—but will return to it, particularly in Chapter 8. We also have a natural way of modelling normic support comparisons: a body of evidence E normically supports a proposition P *more strongly* than it normically supports a proposition Q just in case there is a world at which E is true and Q is false which is more normal than any world at which E is true and P is false.[12]

On my account, in order for one to have justification for believing a proposition P, it is necessary that one's body of evidence E normically support P—it is necessary that all the most normal worlds at which E is true are worlds at which P is true. The probability of P given E can reach any level, short perhaps of 1, without this condition being met. If the probability of P given E is less than 1, there will be possible worlds at which E holds and P does not. If these are amongst the most normal

[11] The idea that possible worlds or states of affairs might be ordered with respect to their normalcy has been proposed before. It has been explored in connection with conditional logics for defeasible reasoning (Delgrande 1987, Boutilier 1994, Boutilier and Becher 1995), in connection with counterfactual conditionals (Gundersen 2004), and in connection with conditional analyses of causation (Menzies 2004). I shall return to some of these applications of the idea in Chapter 6.

[12] It is a consequence of this definition that E normically supports P iff E normically supports P more strongly than ~P. A corresponding principle for justification might be thought counterintuitive: one has justification for believing P iff one has more justification for believing P than ~P. More generally, given this definition of comparative normic support, it will turn out that, if there is a proposition Q such that E normically supports P more strongly than Q, then E must normically support P. Once again, the corresponding principle for justification seems questionable: if there is a proposition Q such that one has more justification for believing P than Q, then one has justification for believing P. Since normic support is proposed here merely as a *necessary* condition on justification, these latter principles won't be automatically derivable from the former principles. I will have more to say about these issues in Chapter 5.

worlds at which E holds, then E will not normically support P. On my account, the probability of P given E can reach any level, short perhaps of 1, without one having justification for believing P.[13]

One consequence of choosing to model normic support in terms of possible worlds is that it will make the notion too coarse to discriminate between propositions that share the same possible worlds truth conditions. Propositions that are true at exactly the same worlds must, on the present model, have the same normic support profile—and, in particular, propositions that are true at all worlds will be normically supported by all bodies of evidence while propositions that are true at no worlds will be normically supported by none. One might regard this as an important objection to the present model—but to dismiss it on these grounds alone would, I think, be hasty.

One thing to note right away is that the notion of evidential probability, as standardly modelled, is *just as coarse*. Probability functions are standardly defined over sets of possible worlds—with the probability of a proposition taken to be determined by the probability of the set of worlds at which it's true (see, for instance, Williamson 2000: s. 10.4, Hájek 2003, Douven and Williamson 2006). As such, propositions that are true at exactly the same possible worlds will receive the same evidential probability—and, in particular, propositions that are true at all worlds will receive probability 1 relative to any possible body of evidence, while propositions that are true at no worlds will receive probability 0.

[13] I am inclined to think that there will be possible cases in which one lacks justification for believing even propositions that have a probability of 1, given one's evidence. If the underlying probability space is infinite, then the fact that a proposition P has a probability of 1 given evidence E is consistent with there being possible worlds at which E holds and P does not. If these are amongst the most normal worlds at which E holds, E will not normically support P. Suppose my evidence is sufficient to situate the speed of light within a particular interval, but leaves it equally likely that any of the values in this interval represents the true speed (our best evidence regarding the speed of light is, I take it, much like this). For any integer n, this interval could be divided into n mutually exclusive and jointly exhaustive sub-intervals such that it is equally likely, given my evidence, that the true value lies within any one of these sub-intervals as any other. The evidential probability that the speed of light is any particular value x within the interval must, then, be less than 1/n for any integer n—that is to say, it must be equal to 0, unless we are willing to countenance infinitesimals. Although the probability that the speed of light is not equal to x has an evidential probability of 1, this is perfectly consistent with there being possible worlds, consistent with my evidence, in which the speed of light is equal to x. If these worlds lie amongst the most normal worlds that are consistent with my evidence, then my evidence will not normically support the proposition that the speed of light is not equal to x.

There are various ways of attempting to make evidential probabilities more discriminating—and these options are available in the case of normic support as well. I'll have more to say about this in section 6.2. In my view, though, issues of this general kind need to be understood as by-products of attempting to treat notions like normic support or evidential probability in a formal way. In building formal models of normic support or of evidential probability, we strike a familiar bargain: we secure a kind of systematicity at the cost of a certain fineness of grain that may appear to be present in the informal notion.

There is, of course, no obligation to offer any formal model of normic support—we might work exclusively with the informal notion, defined in terms of explanation and the associated notion of normalcy. One of the benefits of a formal model, though, is that it allows us to definitively settle questions of logic. It will turn out, for instance, that normic support, as modelled here, is closed under multiple premise deductive consequence—if a body of evidence normically supports each of a set of propositions, then it will also normically support any proposition that they jointly entail. If evidence E normically supports a proposition P, then the most normal E-worlds are P-worlds. If evidence E normically supports a proposition Q, then the most normal E-worlds are Q-worlds. If P and Q jointly entail R then all $P \wedge Q$-worlds are R-worlds. It follows straightforwardly that all the most normal E-worlds are R-worlds. The reasoning can be easily generalised for any number of premises.

It's worth noting that the informal characterisation of normic support already made it very plausible that it should possess this property. If the falsity of P would require special explanation given one's evidence and the falsity of Q would require special explanation given one's evidence, then presumably the falsity of $P \wedge Q$ would also require special explanation, given one's evidence. After all, the falsity of $P \wedge Q$ must involve either the falsity of P or the falsity of Q. The formal model, however, allows us to set this impression in stone.

Thus far, normic support has been proposed only as a necessary condition for justification. As such, the fact that normic support is closed under multiple premise deductive consequence won't guarantee that justification is similarly closed, though it will leave open the path to such a view. Whether we ultimately have this result will depend on what else, if anything, we take justification to require. I shall return to this issue in Chapter 4.

The present model also predicts, relatedly, that no body of evidence could provide simultaneous normic support for each of an inconsistent set of propositions. If E normically supports P and normically supports Q, then P holds at the most normal worlds at which E holds and Q holds at the most normal worlds at which E holds. It follows that there are possible worlds at which P and Q both hold in which case P and Q are consistent. This reasoning can easily be generalised for any number of propositions. Once again, this conclusion rings true even for the informal characterisation of normic support. If the falsity of P would require special explanation given one's evidence and the falsity of Q would require special explanation given one's evidence, then the falsity of $P \wedge Q$ would require special explanation given one's evidence. But the falsity of a contradiction never requires special explanation.

As discussed in Chapter 1, a body of evidence *could* simultaneously make each of an inconsistent set of propositions arbitrarily likely—and it is for this reason that high evidential probability does not normatively coincide with knowledge. Perhaps one could never prove that normic support and knowledge normatively coincide without giving a substantial theory of knowledge—but there is no *formal* obstacle to their doing so. Unlike high evidential probability, normic support has the right logical features in order to normatively coincide with knowledge.

2.5. Objections and Replies

As I suggested in section 2.2 my description of the laptop example, and, in particular, of the evidence available to me, may be contentious. According to the *knowledge account* of evidence (Williamson 2000: chap. 9) one's evidence is equal to one's knowledge—that is, one's body of evidence consists of all and only the propositions that one knows. One who is impressed by this account might object to my description as follows: once I *see* that Bruce's laptop is displaying a blue background, I come to *know* that it is and, according to the knowledge account of evidence, this suffices for the proposition to qualify as part of my body of evidence. My evidence, then, will include not just E_1 and E_2, but P_1 *as well*. In this case, the probability of P_1, given my evidence, will be 1 and, thus, *will* exceed the probability P_2, given my evidence, contrary to what I have claimed.

There is something dissatisfying about the way in which the knowledge account of evidence would have us treat this case—though it's difficult to put one's finger on exactly what it is. Part of what is puzzling about the case is that my belief in P_1 seems as though it could qualify as knowledge, while my belief in P_2 does not, even though my evidence for P_2 seems to be stronger than my evidence for P_1. One way to solve this puzzle is to argue that my evidence for P_1 really is stronger than it appears. According to the knowledge account, P_1 is indeed more strongly supported by my evidence than P_2—but this is only *because* my belief in P_1 is taken to qualify as knowledge. It's difficult to shake the impression that this is a kind of sleight of hand—and that *nothing* has been explained at all.

It may be that these remarks just betray an unwillingness to enter into the spirit of 'knowledge-first' epistemology—but we needn't pursue this matter further here. In order to answer this objection, we need only point out that the force of the laptop example does not depend in any way upon the *actual background colour*—either of Bruce's laptop or of mine. Indeed, my original description of the case left it open what colour background the two laptops were actually displaying. Suppose that the background colour of both laptops is, in fact, red and I *really do* hallucinate when I look at Bruce's (unbeknownst to me of course). In this case, even the proponent of the knowledge account would, presumably, have to restrict my relevant evidence to something along the lines of E_1 (see Williamson 2000: 198). But the relevant intuitions are unchanged. Even though I'm now wrong about the background colour of both laptops, it still seems that I have *better* justification for believing that Bruce's background is blue than I do for believing that my background is blue. It would still be more legitimate for me to assert or act upon the former proposition than the latter. And the former belief is still a *candidate* for knowledge, while the latter is not.[14]

[14] According to epistemological disjunctivists (McDowell 1982, Pritchard 2012), there is a fundamental epistemic asymmetry between veridical and non-veridical perceptions, such that the former put one in a kind of 'direct contact' with a fact in the world, while the latter do not. Disjunctivists may also take exception to my original description of the laptop example, arguing that P_1 becomes a part of my evidence when I veridically perceive that Bruce's laptop is displaying a blue background. Once again, this objection can be answered by pointing out that it is in no way essential to the example that my perceptual experience of Bruce's laptop even *be* veridical—we could perfectly well suppose it to be a hallucination. It is also important to point out that examples of this sort need not involve a contrast between statistical and perceptual evidence—as the Blue-Bus case illustrates, we can, for instance, build a corresponding example around a contrast between statistical and testimonial evidence.

The knowledge account can, in fact, be made to generate the same evidential predictions without our needing to suppose that I'm mistaken about the background colour of Bruce's laptop. We could imagine instead that I'm Gettiered when I set eyes on Bruce's laptop. Suppose there was some appreciable risk that I might have suffered a colour hallucination at that moment, even though this didn't eventuate and my perceptual experience was perfectly veridical. So long as the proponent of the knowledge account buys into the standard verdict about cases like this, he will, as before, have to restrict my evidence to something along the lines of E_1, as this is the only relevant knowledge in the vicinity. But whether one is Gettiered with respect to a proposition should have no effect on whether one has justification for believing it—or so the conventional thinking would have it, at any rate.

The knowledge account of evidence may not be the most natural fit with my original description of the evidence available to me in the laptop example—but, as this discussion shows, it is not at all inconsistent with this description. On the contrary, provided the details of the example are filled out in an appropriate way, the knowledge account can be made to explicitly sanction the claim that my relevant evidence consists of just E_1 and E_2.

While a proponent of the knowledge account of evidence might worry that I'm underestimating the evidence that's available to me, others may be concerned that I'm *overestimating* this evidence. Perhaps E_2 is not the kind of proposition that could literally be a part of my evidence. If my laptop is working and is properly implementing the algorithm, then it will be 99.9999% likely to be displaying a blue background—but I could perhaps have some doubts as to whether my laptop is working and is properly implementing the algorithm.

Let Pr be my evidential probability function at the point I set eyes on Bruce's laptop. Suppose that E_2 is not a part of my evidence but itself has some evidential probability that is close to but less than 1. This will make the evidential probability of P_2 more difficult to calculate, but it need not make any difference to the overall thrust of the example. If $0 < Pr(E_2) < 1$ then, by the theorem of total probability, $Pr(P_2) = Pr(P_2 \mid E_2).Pr(E_2) + Pr(P_2 \mid {\sim}E_2).Pr({\sim}E_2)$. $Pr(P_2 \mid E_2)$ is, of course, equal to 0.999999. $Pr(P_2 \mid {\sim}E_2)$ will be more difficult to determine, but it will take some non-zero value. Let x be the value of $Pr(P_1)$—the probability, given my evidence, that Bruce's laptop is displaying a blue background. Provided, then, that $Pr(E_2) > x/0.999999$,

P_2 will turn out to be more likely, given my evidence, than P_1—as the example requires. This result holds irrespective of the value of $Pr(P_2 \mid {\sim}E_2)$.

More generally, if we let $Pr(P_2 \mid {\sim}E_2) = y < 0.999999$, it is enough that $Pr(E_2) > (x - y)/(0.999999 - y)$.

Proof

$$Pr(E_2) > (x - y)/(0.999999 - y)$$

$$\Rightarrow \; Pr(E_2).(0.999999 - y) > x - y$$

$$\Rightarrow \; Pr(E_2).0.999999 - Pr(E_2).y > x - y$$

$$\Rightarrow \; Pr(E_2).0.999999 + y - Pr(E_2).y > x$$

$$\Rightarrow \; Pr(E_2).0.999999 + (1 - Pr(E_2)).y > x$$

$$\Rightarrow \; Pr(E_2).0.999999 + (Pr({\sim}E_2)).y > x$$

$$\Rightarrow \; Pr(P_2 \mid E_2).Pr(E_2) + Pr(P_2 \mid {\sim}E_2).Pr({\sim}E_2) > Pr(P_1)$$

$$\Rightarrow \; Pr(P_2) > Pr(P_1)$$

QED

If, for instance, $Pr(P_1) = 0.95$ and $Pr(P_2 \mid {\sim}E_2) = 0.4$ then, provided that $Pr(E_2)$ is greater than 0.92, P_2 will be more likely, given my evidence, than P_1. In order for this example to serve its purpose, then, it is not necessary that E_2 be a part of my evidence or even that it be certain given my evidence. Rather, it is enough that E_2 meet a given evidential probability threshold, which can be determined as a function of $Pr(P_1)$ and $Pr(P_2 \mid {\sim}E_2)$.

Before concluding, it's worth noting that there are further examples, of the same apparent structure as the laptop example, that don't seem to rest on any assumptions about a person's evidence. The following is adapted from an example given by Don Fallis (1997): suppose I'm presented with two very large numbers n_1 and n_2 and asked to determine, within a certain time limit, whether or not they are prime. I know of one sure-fire method for detecting primeness—but the method is cumbersome and difficult. I apply this method to n_1 and, after carefully checking through all of my calculations, arrive at the verdict that it is indeed prime. I then notice that most of my time has elapsed so resort to a different method for n_2. This method is much quicker and easier, but it is not sure-fire—it may yield the verdict that a number is prime even if it is in fact composite, though it is very *unlikely* to do so. I apply the method to n_2 and arrive at the verdict that it is prime.

Let's fill in a few further details: it has been proven that, if a number n is composite, then the majority of the natural numbers between 1 and n will stand in a certain, easily detected, numerical relationship to n, whereas if n is prime, then none of the natural numbers between 1 and n will stand in this relation to n (Rabin 1980). Suppose that, in the time I have left, I manage to test a sample of numbers less than n_2 and find that none has the relation in question to n_2. If my sample is sufficiently large, then the chance of this result, given that n_2 really is composite, may be very low.

Under these circumstances, I would seem to have more justification for believing that n_1 is prime than I do for believing that n_2 is prime. Indeed, it seems I would not be justified in believing outright that n_2 is prime. A number of further observations accompany this: if I announce, once the time runs out, that both numbers are prime then, with respect to n_2, what I seem to be doing is making an educated *guess* (which can of course be a perfectly reasonable thing to do in a test). When it comes to n_1, however, it wouldn't be at all natural to describe my announcement as a 'guess', educated or otherwise. And, once I'm outside of the test context, it seems that I'm still perfectly entitled to announce that n_1 is prime, but I shouldn't simply announce the same thing about n_2—the most that I should say, it seems, is that it is overwhelmingly likely to be prime.

As noted, though, the first method for detecting primeness is much more difficult to apply than the second and, in spite of my careful checking, the probability that I've made an error in applying the first method will be higher than the probability that I've made an error in applying the second method. Once these probabilities of error are factored in, the overall probability that the first number is prime could very well be lower than the overall probability that the second number is prime. If pressed about my own fallibility I might even judge the second number to be more *likely* prime than the first, and I might be quite right to do so.[15] This judgement is not based, however, on any assumptions about what my

[15] Fallis (1997) describes this kind of example in the course of arguing that there is no epistemic basis for the current mathematical practice of rejecting probabilistic methods (such as the method described here) as a means of establishing mathematical truths. Against the background of a risk minimisation picture, this conclusion may be correct—a probabilistic method with a low rate of implementation error can minimise risk more effectively than a deductive method with a high rate of implementation error. The notion of normic support could however offer one way of legitimising the current practice. I won't pursue this further here. For discussion of Fallis's position, and of this particular example, see Easwaran (2009).

evidence consists of—I'm simply estimating the error risks associated with these methods themselves along with the error risks associated with my attempting to implement these methods. It's not at all clear what we should consider to be my relevant evidence in this case—arguably, the very notion of evidence seems of limited use in helping us to understand this sort of epistemic predicament.

In this chapter I've argued that the risk minimisation conception of justification is flawed and have begun to sketch an alternative picture. In section 2.1, I quoted some advice from Descartes—namely, that one should never believe that which is merely probable. In one sense, I concur with this. If the *only* thing that can be said in favour of a proposition is that it is probable then, in my view, one would not be justified in believing it. In another sense, I, like any fallibilist, will reject Descartes's advice—in my view one can be perfectly justified in believing things that are less than *certain*. The compatibility of these two views owes to the fact that there are ways in which a proposition can be more than probable, though less than certain. I have outlined one such way here.

3

Justification and Lotteries

3.1. The Lottery Paradox

Lotteries, and our intuitions about them, lie at the heart of a number of persistent puzzles in epistemology. In this chapter, I shall think through some of these puzzles, in light of the theory of justification just sketched. Perhaps the most famous and enduring lottery-related puzzle in epistemology is the so-called lottery paradox described by Kyburg in *Probability and the Logic of Rational Belief* (1961) and briefly mentioned in section 1.3. Suppose I hold a single ticket—ticket #542—in a lottery that I know to have a thousand tickets and one guaranteed winner. Suppose the lottery has been drawn but I'm yet to hear anything about the result. It seems plausible, at first, that I have justification for believing that my ticket has lost. But, given that all the tickets are on a par, and there is nothing special about ticket #542, it would seem to follow that I have justification for believing the same thing about each of the other tickets—justification for believing that ticket #1 has lost, that ticket #2 has lost etc. But these 1,000 'lottery propositions' together entail that *all* of the tickets have lost, in which case, given multiple premise closure, this is also something that I must have justification for believing. By stipulation, I have justification for believing that some ticket has won in which case, by a further application of multiple premise closure, I must have justification for believing an out and out contradiction; that every ticket has lost the lottery and that one of the tickets has won the lottery.

Kyburg opted to resolve the paradox by rejecting the principle of multiple premise closure (Kyburg 1961, 1970)—and many have followed him in this (see, for instance, Schick 1966, Derksen 1978, Foley 1993: chap. 4, 2009, Hawthorne and Bovens 1999, Sturgeon 2008). And anyone who holds to the *letter* of the risk minimisation conception, as set out in Chapter 2, is effectively committed to resolving the paradox in this way. However, in spite of the pre-eminence of the risk minimisation conception, the idea that justification is closed under deductive inference—even of a multiple premise kind—is one that many epistemologists hold dear.[1]

Many epistemologists, when confronted with the lottery paradox, have attempted to preserve multiple premise closure and to motivate instead the denial of the first premise—that I have justification for believing that ticket #542 has lost. There has, in fact, been a certain tradition of attempting to engineer this result by refining the basic risk minimisation conception of justification. On these 'refined' risk minimisation theories, high evidential probability remains as a necessary condition for justification, but only suffices for justification when further special probabilistic conditions are satisfied. I will discuss these theories in some detail in Chapter 9—for now, the following proposal, due to Lehrer (1974: chap. 8), should suffice to give the general flavour:

> One has justification for believing a proposition P if it has an evidential probability that exceeds the threshold t and, in addition, exceeds the evidential probability of any competitor proposition, where a proposition Q is a competitor to P just in case the evidential probability of P is lower on the assumption that Q than otherwise.

Lehrer's theory does seem to give the desired result—at least in the simple lottery case described. Since the lottery is guaranteed to have a winner, the propositions that ticket #1 has lost, that ticket #2 has lost . . . that ticket #1,000 has lost, are competitors of one another. Since the lottery is fair, each of these lottery propositions is equally likely. As such, Lehrer's theory, unlike the basic risk minimisation theory, does not grant justification for believing any one of them. Lehrer's theory clearly preserves the spirit of risk minimisation, and enables us to resolve the lottery paradox (at least as it stands) without sacrificing closure. As I will discuss in Chapter 9, Lehrer's proposal, and others that are like it, turn out to be deeply problematic on closer inspection.

[1] The idea that deduction represents an epistemically safe way of extending one's corpus of beliefs was a central component of the Cartesian project in epistemology—a project that still, undoubtedly, exerts a profound influence on epistemic thinking. For Descartes, though, this idea went hand in hand with another that has fared less well historically—namely, the infallibilist idea that justification requires the complete elimination of error risk. At the very least, what puzzles such as the lottery paradox show us is the way in which closure becomes a very different sort of prospect once infallibilism is abandoned.

My inclination, of course, is also to reject the first premise of the lottery paradox—but not to try and reconcile its rejection within a risk minimisation framework. As Jonathan Vogel observes:

> although winning a lottery on a particular ticket is unlikely or improbable, it would not be *abnormal* in some intuitive sense, for the ticket one holds to turn out to be a winner.
>
> (Vogel 1990: 16)

I have claimed that, in order for one to have justification for believing a proposition P, one's body of evidence must normically support P. My only relevant evidence in the lottery case is the fact that the lottery is fair and there are one thousand tickets and only one winner. This evidence does not normically support, for any particular ticket, the proposition that it has lost—the one thousand possible lottery outcomes will all be on an explanatory par. Amongst the most normal worlds in which this evidence holds true will be worlds in which ticket #1 wins, worlds in which ticket #2 wins, worlds in which ticket #3 wins, etc. Thus, I lack justification for believing, of any particular ticket, that it has lost.

One reason that epistemologists have traditionally been squeamish about denying that one can justifiably believe lottery propositions is for fear of inviting radical scepticism. If we don't have justification for believing, of a single ticket, that it has lost, then surely we don't have justification for believing much at all. Richard Foley clearly expresses this worry:

> most of us, on reflection, would be likely to regard the evidence we have for the claim that ticket one will not win to be as strong as the evidence we have for most of the claims that we believe. For instance, we would be likely to regard the evidence for it as being at least as strong as the evidence we have for the claim that the room we have just left still has furniture in it . . .
>
> (Foley 1987: 245)

My response to Foley's concern is, perhaps, easy to anticipate: there is more than one notion of evidential strength. Relative to one notion, what Foley writes may be true. Relative to another, it is false.

Suppose that my relevant body of evidence E consists of the proposition that I saw furniture in the room 30 seconds ago and that a fair, billion ticket lottery with a single guaranteed winner has just been drawn. Let F be the proposition that the room has furniture in it now and L be the proposition that ticket #1 has lost the lottery. F may be no more probable, given E, than L is—indeed L may well be more probable. Foley supposes, however, that this is all that there is to say about the bearing of the evidence—and this is a mistake. E normically supports F but does not normically support L. If E were true and F false, there would need to be some explanation for this—such as the floor collapsing or very fast-working burglars or the interventions of a mischievous demon. While E *probabilistically* supports L more strongly than F, it *normically* supports F more strongly than L.

3.2. Sceptical Threats

Denying that we have justification for believing lottery propositions, as we've just seen, involves a potential sceptical threat. Lottery propositions can have any evidential probability that we like, short of 1. As such, for any proposition that is less than certain, we can always imagine a lottery proposition that is more evidentially likely to be true. The only further assumption that we need, then, in order to obtain a general sceptical result, is this: if I lack justification for believing a proposition P and P is more evidentially likely, for me, than a proposition Q, then I also lack justification for believing Q. If we lack justification for believing lottery propositions, it follows from this assumption that we lack justification for believing anything that is less than certain.

This assumption has some intuitive appeal, but is far from innocuous—it is, in effect, an alternative formulation of thesis (a) of the risk minimisation conception as set out in section 2.1.[2] The key to resolving this sceptical problem, in my view, is to free ourselves from this conception. But even if this immediate sceptical problem can be defused, the idea that we lack justification for believing lottery propositions may contain a further, and somewhat subtler, sceptical threat.

As John Hawthorne observes in *Knowledge and Lotteries* (2003), lottery propositions appear to be entailed by a range of ordinary, mundane propositions that we believe (see also Vogel 1990). Suppose I've

[2] That thesis (a) entails this assumption is obvious. The reverse direction takes a little more work to see. What the assumption amounts to, put a little more formally, is that justification is a *monotonic* function of evidential probability—the set of propositions justified by a body of evidence must be closed under the relation of greater-than-or-equal evidential probability. Relative to a given body of evidence, then, we can always identify an evidential probability threshold that is both necessary and sufficient for justification—namely, the greatest lower bound of the set of evidential probabilities assigned to the justified propositions.

made a lunch appointment with a friend for tomorrow and suppose I know my friend to be very reliable about such things. Under these circumstances, it's plausible that I have justification for believing that I'll be having lunch with my friend tomorrow. Presumably, though, I have very good reason to think that, if my friend were to win an enormous cash prize in a lottery tomorrow, she would either cancel or miss our lunch. But these two propositions deductively entail that my friend won't win a lottery tomorrow. If I lack justification for believing this conclusion then, given multiple premise closure, I must lack justification for believing one of the premises—and the natural candidate is the proposition that I'll be having lunch with my friend tomorrow.

In thinking through this example, though, we need to proceed with some caution. Believing that someone won't win a lottery *when I have no reason to think that she even holds a ticket in a lottery* is a very different prospect from believing that someone won't win a lottery when I know that she holds a ticket in a lottery and have further information to the effect that the lottery is fair, etc. My theory of justification predicts that no belief of the latter sort could be justified—but it makes no definitive predictions about beliefs of the former sort. In the example under consideration, it's consistent to think that in the most normal worlds in which my evidence holds—in which I've made the lunch appointment, my friend is reliable, etc.—she *doesn't* hold a ticket in a lottery to be drawn tomorrow and, thus, doesn't win any such lottery. If this is right, then the proposition that my friend will not win a lottery tomorrow will be normically supported by my evidence. The proposition that I'll be having lunch with my friend tomorrow may also be normically supported and could be something that I have justification for believing.

We could, of course, alter the case so that I *am* aware of a fair lottery to be drawn tomorrow and aware that my friend holds one or more tickets in this lottery. If my evidence were expanded to include these propositions then, quite clearly, it would no longer normically support the proposition that I'll be having lunch with my friend tomorrow. In some of the most normal worlds in which my friend holds a ticket in such a lottery (and in which we made the lunch appointment and she is reliable), her ticket wins and she cancels or misses our lunch. In this case, according to my theory, I would indeed lack justification for believing that I'll be having lunch with my friend tomorrow—the most I would be justified in believing is that I will *very likely* be having lunch with my friend tomorrow. But this, I suggest, is not a counterintuitive result—or, at least, not obviously so.

To consider another of Hawthorne's examples, suppose I'm weighing up various possible holiday destinations for next year and, on the basis of my income, mortgage, other expenses, etc., I calculate that I won't be able to afford an African safari holiday. But the proposition that I won't be able to afford an African safari holiday entails that I won't win a large cash prize in a lottery. If I lack justification for believing the latter proposition then, given closure—and it looks as though even *single premise* closure may be enough here if we pad out the propositions a bit—I must also lack justification for believing the former proposition. Once again, if we fill in the details of the case in the right way—I actually own lottery tickets or intend to purchase lottery tickets etc.—my account will predict that I lack justification for believing that I won't be able to afford an African safari holiday and, at best, have justification for believing that this is likely. Once again, though, if we explicitly fill in the details in this way, I don't find this to be a clearly counterintuitive result.

Even if one dislikes the mildly sceptical predictions that my account offers in these cases, it would appear that such results are, at least, relatively quarantined. It is unusual for a belief to entail, given one's evidence, a proposition that literally concerns the outcome of a lottery. The preceding cases needed to be tailored to quite an extent in order to ensure this. According to Hawthorne, though, even if it is unusual for beliefs to entail strict lottery propositions, is not at all unusual for beliefs to entail propositions that *seem very similar*—propositions that are 'lottery-like'. As he writes:

> Just as I have excellent statistical grounds for supposing that any given lottery ticket will lose, I have excellent statistical grounds for supposing that a given apparently healthy person will not have a fatal heart attack very soon. Just as there was no special reason in advance for supposing that the winning ticket was going to win, there was no reason in advance for expecting the worst for some heart attack victim who was apparently healthy. And just as many of our ordinary commitments entail that this or that person will lose a lottery, many of our ordinary commitments entail that this or that person will not soon suffer a fatal heart attack.
>
> (Hawthorne 2003: 3)

If one cannot have justification for believing, in the absence of special, non-statistical reasons, that one will lose a fair lottery, then it seems as though one cannot have justification for believing, in the absence of special, non-statistical reasons, a range of relevantly similar things. As well as the proposition that a young, apparently healthy individual will not soon suffer a fatal heart attack, Hawthorne provides, as examples, the proposition that the president has not died within the last five minutes, that there hasn't been an electrical outage in my neighbourhood, and that the scores in the newspaper are not in error (see Hawthorne 2003: 4, see also Vogel 1990: s. V).

It may be exceedingly unlikely for a young, fit, and apparently healthy person to suffer a fatal heart attack—but it does happen to some people who meet this description. Isn't this relevantly like a lottery then, with all the young, apparently healthy individuals holding 'heart attack' tickets, some of which will inevitably be drawn? Similarly, it seems inevitable that some people have unexpectedly died in the last five minutes, that some neighbourhoods occasionally have power outages, and that newspapers occasionally contain errors. Whether it's *the president* who has died or *my neighbourhood* that currently has an electrical outage or *the scores in the newspaper in front of me* that happen to be in error just depends, surely, on 'who's numbers have come up'.

And Hawthorne seems right to suggest that many of the things we ordinarily believe do entail these 'lottery-like' propositions—almost everything that I believe about a person's future, for instance, will entail that he will not soon suffer a fatal heart attack. As such, if we lack justification for believing all such propositions then, given an appropriate closure principle, a more pervasive scepticism does indeed threaten. As suggested above, it may be that even single premise closure suffices for this.

My response to this sceptical problem is perhaps not as definitive as one might wish—but I want to suggest that the analogy between genuine lottery propositions and lottery-like propositions is simply not as close as Hawthorne suggests. Just because we have statistical evidence bearing upon a proposition, that is not to say that we do, or should, think of its truth or falsity as the outcome of some random process. Undoubtedly, there is a (somewhat pessimistic) mindset in which we might think of future health, mortality, the reliability of the electricity supply, the veracity of newspapers, etc. as tethered to pure chance—and it's not unusual to see lottery-type metaphors used in relation to such things on occasion. But these *are* just metaphors—for the most part our attitude to heart attacks, unexpected deaths, power outages, and newspaper errors is not at all like our attitude to chance events.

If a young, fit, apparently healthy friend were to suffer a fatal heart attack, this is something that would immediately prompt us to seek possible *explanations*: was he suffering from a virus? Did he have a congenital heart abnormality? Was he poisoned? We may even develop a psychological *need* to identify some explanation of this sort. This is not a reaction to something we regard as a 'pure chance' event. If a friend had won the lottery, rather than suffering a fatal heart attack, then, while this might be equally surprising, it would not prompt us to look for explanations. In this case, we'd be perfectly content with the thought that he played the odds and his ticket came up. Such a thought would hardly content us in the heart attack case.

Similar remarks apply to Hawthorne's other examples of lottery-like propositions. If the president has died within in the last five minutes, or there is a power outage in my neighbourhood, or the scores in the newspaper are in error, there would have to be some explanation as to how such things came about. We don't just take a 'you win some you lose some' attitude to events like these. This is a significant disanalogy with genuine lottery propositions. Recall Vogel's claim above: it would not be abnormal, in some intuitive sense, for my ticket to win a fair lottery, no matter how unlikely that outcome may be. And yet it *would be* abnormal, in just this sense, for an apparently healthy individual to drop dead of a heart attack, for the president to have died within the last five minutes, for there to be a power outage in my neighbourhood, or for the scores in the newspaper to be in error.

Given my account of justification, this disanalogy is all important. My evidence will not normically support the proposition that a particular ticket will lose a fair lottery. But, so long as my evidence includes information about the youth and apparent health of an individual, the relative health and safety of the president, the apparent reliability of the electricity supply, and the general veracity of newspapers, it can normically support the propositions that the individual won't soon suffer a fatal heart attack, that the president hasn't died within the last five minutes, that there hasn't been a power outage in my neighbourhood, and that the scores in the newspaper are not in error. My account of justification does indeed offer a way of cleaving apart lottery propositions and lottery-like propositions—a scepticism about the former need not overflow into a scepticism about the latter.[3]

[3] We can, of course, imagine possible events for which it just isn't *clear-cut* whether or not their occurrence would demand explanation—whether or not their occurrence would

3.3. Harman's Puzzle

I turn now to another lottery-related epistemic puzzle—a puzzle that was, to the best of my knowledge, first noted by Harman (1968) but has since been discussed by Dretske (1971), Cohen (1988: s. IV), and DeRose (1996). Suppose that Bob has bought a single ticket in two lotteries, both of which are guaranteed to have a single winner. The first of these is a relatively small community lottery in which just one hundred tickets were sold and the other is a nationwide lottery with a total of one million tickets. Suppose that both of these lotteries have now been drawn and Bob has not won either. Bob watches the drawing of the community lottery on TV and hears the announcer read out the number of the winning ticket which, naturally, is not his. The broadcast of the nationwide lottery draw, however, has been delayed and Bob has not yet seen it.

In seems clear, in this situation, that Bob *knows* that he has lost the community lottery, but does not yet know that he has lost the nationwide lottery. A number of further intuitions seem to accompany this: it seems reasonable for Bob to assert that he has lost the community lottery, but premature for him to assert that he has lost the nationwide lottery. If asked about the nationwide lottery, Bob ought to say something like 'It is *very likely* that I've lost' or 'I've *almost certainly* lost'. This sort of qualification is not required when it comes to the community lottery. It also seems reasonable for Bob to throw away his community lottery ticket stub or to sell it for a cent (perhaps to someone who has not yet

count as genuinely abnormal or as merely unlikely. For such events, it may also be unclear whether we have normic support for the proposition that they won't occur and, in turn, whether we have justification for believing this, and for believing other propositions that entail it. If a tennis player with a world ranking of 200 were to win Wimbledon, would this require special explanation? Perhaps. If so, what about the player ranked 199 and the player ranked 198 and so on? The answers are not all clear-cut. But this would seem, in any case, to track a corresponding unclarity in our judgements about what we have justification for believing. Would I be justified in believing that the player ranked 200 won't win or merely that the player ranked 200 is *very unlikely* to win? If the former, then what about the player ranked 199, and the player ranked 198 etc.? If I've placed a substantial bet on one of these players it may also be unclear whether I have justification for believing that I won't be able to afford an African safari holiday etc. No theory of justification, of course, will offer a clear verdict in every possible case we can describe—and nor should it. Very often what we want from a theory of justification is not a 'resolution' of controversial cases, but an understanding of why they are controversial.

seen the draw). But it would certainly not be reasonable for Bob to throw out his *nationwide* lottery ticket stub or to sell that for a cent.[4]

In spite of all of this, it is very plausible to think that the probability, given Bob's evidence, that he has lost the community lottery is actually *lower* than the probability, given that same evidence, that he has lost the nationwide lottery. After all, what Bob sees on TV does not guarantee that he has lost the community lottery. The person drawing the lottery could, for instance, have misread the number on the ticket. Such things may not happen often but presumably the likelihood of such an occurrence would be somewhat higher than one in 10,000. How is it that Bob can know that he has lost the community lottery, reasonably assert that he has lost the community lottery, and even *reasonably tear up his ticket stub* when the probability, for him, that he has lost the community lottery is actually lower than the probability that he has lost the nationwide lottery?

Perhaps the most common putative solution to this problem exploits the notion of *sensitivity* (Dretske 1971, DeRose 1996). If Bob had won the community lottery then he *would* have heard his ticket number read out on TV. That is, if Bob had won the community lottery, his evidence would have been different and he wouldn't have believed that he lost. In the terminology introduced in section 1.3, we might say that Bob's belief that he has lost the community lottery is based upon *sensitive* evidence and is a sensitive belief. In contrast, if Bob had won the nationwide lottery, he wouldn't have seen or heard anything different—his evidence would have been exactly the same. If Bob were to believe that he has lost the nationwide lottery, his belief would not be sensitive.

This is a striking observation—but it cannot be the full story as to what is driving our intuitions here. Imagine now a slightly different case in

[4] Here is one thing that could be guiding our judgements here: one might assume (though it wasn't specified) that the nationwide lottery promises a much larger *prize* than the community lottery, and this could give us extra reason to think that tearing up the nationwide lottery ticket or selling it for a cent is a bad idea. There may be something to this—but the assumed size of the prize is not I think what's behind our judgements about the case. Suppose Bob is part of a large lottery syndicate—so he and a number of others have entered into an agreement in which each buys a ticket in the nationwide lottery and any winnings are shared out. If the syndicate is large enough, then what Bob stands to personally gain in the event that his ticket wins might be suitably modest. But this doesn't change our judgement—it would still be unreasonable for Bob to just give away his share in the ticket or sell it for a cent, in advance of hearing the lottery result.

which Bob *has* won the community lottery, but the person drawing the lottery actually *does* misread the ticket and announces the wrong number at the televised draw, prompting Bob to form a mistaken belief. It is *just* as reasonable, in this case, for Bob to assert that he has lost the community lottery. It is just as reasonable for Bob to throw away his community lottery ticket stub or to sell it for a cent. As tragic as such actions would be, we could I think *forgive* Bob for taking them. And if he did act in this way, he may well have a legitimate claim to compensation of some kind when the truth emerges. Bob, of course, does not *know* in this case that he has lost the community lottery, but his belief would appear to be a very good candidate for knowledge.

The notion of sensitivity, needless to say, is no help here. Bob's evidence, in this new case, is clearly *not* sensitive to his having lost the community lottery—after all he has actually *won* the community lottery. The notion of sensitivity is not, then, the key to resolving the original puzzle. The fact that Bob's evidence is sensitive to the fact that he has lost the community lottery is an *incidental* feature of the puzzle, and not an essential one.

The new mistaken belief case also shows us that the knowledge account of evidence cannot offer a satisfactory resolution of the original puzzle. With respect to the original true belief case, the proponent of the knowledge account could protest that the proposition that Bob has lost the community lottery is part of Bob's evidence and thus more likely, given that evidence, than the proposition that he has lost the nationwide lottery. But this cannot be maintained with respect to the mistaken belief case, in which even the proponent of the knowledge account will have to admit that Bob's having lost the nationwide lottery is more likely, given his evidence, than his having lost the community lottery.

Enoch, Fisher, and Spectre (2012) exploit the notion of sensitivity, in a very similar way to Dretske and DeRose, in order to explain the preference for testimonial over statistical evidence in the law. In the Blue-Bus case described in section 2.2, we contrasted two scenarios. In both of these, a bus was involved in an incident in which there was some injury and/or property damage. In the first scenario, an eyewitness testifies that the bus involved was owned by the Blue-Bus company. In the second scenario, there is no eyewitness, but there is unusually strong statistical evidence to the effect that 95% of the buses operating in the relevant area were Blue-Bus buses.

Suppose a court does find the Blue-Bus company liable on the basis of this statistical evidence. While this evidence makes it highly probable that the bus involved was a Blue-Bus bus, it could not, it seems, be sensitive to this proposition. If the bus had not been a Blue-Bus bus, 95% of the buses operating in the area would still have been Blue-Bus buses and the court would still have found the Blue-Bus company liable. The most similar possible worlds in which the bus involved was not a Blue-Bus bus are worlds in which the Blue-Bus company has the same market share and the court verdict comes out the same.

Now suppose instead that a court finds the Blue-Bus company liable on the basis of eyewitness testimony. While it may not be as probabilistically strong as the statistical evidence, this testimonial evidence could be sensitive to the proposition that the bus involved was a Blue-Bus bus. If the bus involved had not been a Blue-Bus bus, then the eyewitness would not have said that it was and the court would not have found the Blue-Bus company liable—or so, at any rate, we are invited to suppose. According to Enoch, Fisher, and Spectre, it is the *insensitivity* of the statistical evidence that explains our apprehensiveness about it—our sense that it would be unacceptable for a court to base a finding of liability upon it.

While it's true, however, that the testimonial evidence in the Blue-Bus case *could* be sensitive to the proposition that the bus involved was a Blue-Bus bus, whether it really *is* sensitive depends on further features of the case—features that have yet to be spelled out. In the original description of the case, it was never specified, for instance, that the witness testimony was truthful and that the bus involved really was a Blue-Bus bus. Let's suppose instead that the witness testimony is false and that the bus involved was owned by another company. In this case, the testimony is not sensitive to the proposition that the bus involved was a Blue-Bus bus. There is a very similar possible world—namely, the *actual* world—in which the bus involved was not a Blue-Bus bus, even though the witness testified that it was. Provided, though, that the testimony is not in any way challenged or called into question in court, it would still be acceptable for the Blue-Bus company to be found liable. Arguably, this is the verdict that the court would be *obliged* to reach.[5] Whatever it is that is driving our intuitions about acceptability, it is not facts about sensitivity.

[5] A similar criticism applies to Thompson's attempt to vindicate the legal difference between testimonial and statistical evidence using a causal condition (see Thompson 1986).

The notion of normic support offers an alternative resolution of this puzzle. As discussed in section 2.3, if an eyewitness testifies that the bus involved was a Blue-Bus bus, even though it wasn't, then there would have to be some explanation as to how this came about. It can't 'just so happen' that the testimony was wrong—there has to be more to the story. But it could 'just so happen' that the bus involved was not a Blue-Bus bus in spite of the fact that 95% of the buses operating in the area on the day in question were Blue-Bus buses. The testimonial evidence normically supports the conclusion that the bus involved was a Blue-Bus bus while the statistical evidence does not. This holds true irrespective of whether the testimony is correct or incorrect—the crucial point is that it *could not be* incorrect without there being some accompanying explanation. On my account, we have justification for believing that the bus involved was a Blue-Bus bus on the basis of the testimonial evidence, but not on the basis of the statistical evidence.

Turning back to the lottery, if Bob had won the community lottery, in spite of the announcement that he heard on TV, then this would require some explanation—the announcer misread the ticket number or some such. In contrast, if Bob had won the nationwide lottery, in spite of the odds against it, then no special explanation would be needed. Someone's ticket had to win—it might just as well have been Bob's as any other. The one million possible outcomes are on an explanatory par. Bob's evidence, in *both* the original and the mistaken belief cases, normically supports the conclusion that he has lost the community lottery, but does not normically support the conclusion that he has lost the nationwide lottery. As a result, Bob has justification for believing, in both cases, that he has lost the community lottery, but does not have justification for believing that he has lost the nationwide lottery—not, at any rate, until he hears the announcement.[6] This is why it is reasonable for Bob to assert that he

As Thompson observes, in the true testimony case the testimonial evidence is caused by the fact that the bus involved was a Blue-Bus bus, while the statistical evidence is not. But the observation does not, of course, carry over to the case in which the testimony is false.

[6] Suppose Bob does eventually see the delayed telecast of the nationwide lottery draw and hears the announcer read out the number of the winning ticket—#724911 say. It's very plausible, at this point, that Bob has justification for believing that ticket #724911 won the nationwide lottery. Consider the following principle: if one receives testimony to the effect that P is true and the antecedent probability of P is exceedingly low—as low or lower than the probability that the testifier is lying or mistaken—then the testimony cannot furnish one with justification for believing P. The principle can seem somewhat natural on a first

has lost the community lottery, but not that he has lost the nationwide lottery. This is why it is reasonable for Bob to act on the proposition that he has lost the community lottery, but not on the proposition that he has lost the nationwide lottery.[7]

reading—and it is sometimes taken to be one of Hume's premises in his well-known argument that one can never justifiably accept testimony to the effect that a miracle has occurred (see, for instance, Hambourger 1980: ss 1 and 2; the most suggestive passage in Hume 1748 is in s. X, §13).

Bob's situation would seem, though, to provide a clear counterexample to this principle. The announcer is clearly testifying to a proposition with an exceedingly low antecedent probability. The evidential probability, for Bob, that ticket #724911 won the lottery, prior to his hearing the announcement, is 0.000001. As discussed above, the evidential probability of the announcer making an error is plausibly higher than this. But none of this inclines us in the least to deny that Bob acquires justification for believing that #724911 is the winning ticket—to think that he should dismiss the announcement and suspend judgement on this matter etc.

Interestingly, if my way of thinking about justification is on the right track then what *could* prevent testimony from supplying justification is not the prior improbability of its content but, rather, the prior *abnormality* of its content. Put somewhat crudely, the reason that testimony is capable of supplying normic support for its content is that mistakes and lies, in general, make for abnormal states of affairs—departures from ideal normalcy. But the normic support supplied by testimony will lapse in the event that the truth of what is being testified to would represent at least as great a departure from ideal normalcy as would deception or error on the part of the testifier. If normic support is taken to be necessary for justification, then we might endorse the following: if one receives testimony to the effect that P is true and P would be as abnormal or more abnormal than the testifier lying or being mistaken, then the testimony cannot furnish one with justification for believing P. A principle of this kind could perhaps be used, in place of the principle above, in giving an alternative reconstruction of Hume's argument concerning miracles. After all, the proposition that a miracle has occurred, unlike the proposition that a particular ticket won the lottery, is exceedingly *abnormal* (as well as being exceedingly unlikely). See principle (3) in Hambourger (1980) for a related suggestion. I won't pursue this further here.

[7] There may, of course, be other treatments of Harman's puzzle that don't appeal to the notion of normic support. Consider again the mistaken belief case in which Bob's ticket wins the community lottery, but the wrong number is announced. Although Bob's belief that his ticket has lost is not sensitive, we might argue that it *would have been sensitive*, if it had been true—that is, if Bob really had lost the community lottery, as the testimony suggests, then his belief would have been based on evidence that is sensitive to this fact. One might propose that it is this feature, or something like it, that explains our differing judgements about the community and nationwide lotteries. (Proposals of this kind have been suggested to me a number of times informally—see also Enoch, Fisher, and Spectre 2012: n. 23.)

Further cases will prove problematic for this proposal, however. Suppose Bob comes to believe that his ticket has lost the community lottery after hearing a different ticket number announced as the winner. Suppose this is true—Bob's ticket really has lost—but, unbeknownst to Bob or anyone else, the announcer was unable to read the number on the winning ticket and just blurted out the first number that came to mind. It would be perfectly reasonable for Bob to assert or act on the proposition that his ticket has lost. It is not the

The fact that one should not in general assert that one has lost a fair lottery, purely on the basis of the odds involved, has been observed by others—notably Williamson (2000: s. 11.2) and Hawthorne (2003: s. 1.3). Both Williamson and Hawthorne, as mentioned in section 1.3, take this observation to support the *knowledge account* of assertion, according to which one should only assert what one knows to be true. Put simply, their reasoning is something like this: if one should not assert that one has lost a fair lottery, then no evidential probability threshold shy of 1 can suffice for warranted assertion. Since only one's knowledge has an evidential probability of 1, one should only assert what one knows to be true.

The assumption here is clear enough: any appropriate epistemic standard less demanding than knowledge will have to be understood in terms of an evidential probability threshold, shy of 1. Williamson does consider the possibility that evidence may, in some cases, support propositions in a way that does not reduce to probabilification, but then falls back on something very close to the risk minimisation conception, claiming that, if one lacks warrant to assert a proposition, then surely one lacks warrant to assert any proposition that is *less likely* for one to be true (Williamson 2000: 251).[8] Once we give up this principle, we can see that the knowledge account of assertion draws little support from consideration of lottery cases (see Nelkin 2000, Douven 2006: 464, Kvanvig 2009).

3.4. Vogel's Puzzle

Another puzzle concerning lotteries and our intuitions about them was introduced by Jonathan Vogel (1999) and discussed by Hawthorne (2003: s. 1.2): suppose that every year Sue's community hosts a hundred

case, however, that Bob's belief would have been sensitive if true. It *is* true, and it *isn't* sensitive. If Bob's ticket had won, the same number would have been announced and Bob would still have believed that it had lost. In any case, proposals like this one seem largely motivated by a desire to retain *some role* for sensitivity in the treatment of Harman's puzzle—and yet it's difficult to see what rationale there could be for this. The normic support proposal is, in comparison, simple and independently motivated, tying in with a broader theory of epistemic justification.

[8] This is just the assumption used to drive the first sceptical problem described in section 3.2, but with 'warranted assertion' substituted for 'justified belief'. As demonstrated in n. 2, the assumption is equivalent to thesis (a) of the risk minimisation conception. This is not to deny that Williamson's argument has intuitive force—thesis (a) of the risk minimisation conception has intuitive force. The point, rather, is that this way of arguing will be illegitimate in any context in which the risk minimisation conception is in question.

ticket lottery and Sue is in the habit of buying a ticket. As with any lottery case, it's intuitive that I cannot know, purely on the basis of the odds involved, that Sue will not win this year's community lottery and, further, that I cannot reasonably assert or act upon this proposition. But now consider the proposition that Sue is not going to win the community lottery *every year for the next fifty*. This, at first blush, *does* seem like the kind of thing I could reasonably assert or act upon. Indeed, it seems intuitive to many that this is something I can *know to be true*.

This intuition presents a problem for my account of justification. Suppose my relevant evidence consists of the propositions that a community lottery will be held every year for the next fifty and that Sue will hold a ticket in each one of these lotteries. It's clear that this evidence does *not* normically support the proposition that Sue will not win every one of these lotteries. For each of the 10^{100} possible outcomes of these fifty community lotteries, there will be a maximally normal world in which my evidence holds and the outcome obtains. One of these outcomes involves Sue winning every lottery. On my account, I could not have justification for believing that this outcome won't obtain, contrary to intuition.

On close inspection, though, this intuition presents no special problem for my account—rather, it proves problematic *irrespective* of one's background theory of justification. One way to see this is by contrasting this iterated lottery case with a case in which Sue holds a single ticket in a one-off mammoth lottery with 10^{100} tickets. When it comes to this mammoth lottery, our standard lottery intuitions would seem to engage—I cannot know that Sue's ticket won't be the winning ticket and I cannot reasonably assert or act upon this. In spite of this, the odds of Sue winning the mammoth lottery and the odds of her winning the community lottery every year for the next fifty are, given a few idealising assumptions, *exactly the same*. Surely our epistemic situation with respect to the outcomes of a sequence of iterated lotteries is no different from our epistemic position with respect to the outcomes of a single mammoth lottery—we have no special insight into the former that we lack with respect to the latter.

We could even imagine that the mechanism whereby the mammoth lottery is drawn involves running a sequence of fifty 100-ticket lotteries with each possible mammoth lottery outcome corresponding to one of the possible fifty ticket sequences. This would, after all, be a perfectly fair, random mechanism (provided the mechanism for each of these 100-ticket lotteries is random and fair). In this case, Sue's winning the mammoth lottery could literally involve the *same sequence of events* as Sue's winning the next fifty community lotteries. In both cases it's just a matter of Sue's ticket sequence (#47, #24, #97 . . . or what have you) being the sequence of winning tickets in the fifty 100-ticket lotteries. How could I possibly have justification for believing that this won't be the winning sequence when considering iterated community lotteries but *not* have justification for believing this when considering a single mammoth lottery? It's difficult to see that there could be any epistemically relevant difference between the iterated lottery and the mammoth lottery that a viable theory of justification could pick up on.

The name of the game, then, is not to try and *accommodate* the contrary intuitions about the iterated lottery and the mammoth lottery—it is, rather, to *explain away* one or other of these intuitions and restore a kind of symmetry. By 'explaining away' an intuition, I simply mean telling some plausible story as to why the intuition seems attractive or compelling, in spite of being mistaken. On the risk minimisation conception of justification, it is the intuition about the iterated lottery that is, in effect, correct, and the intuition about the mammoth lottery that is mistaken and needs explaining away. On my account of justification, it is the intuition about the mammoth lottery that is correct and the intuition about the iterated lottery that is mistaken and needs explaining away.

The first thing to notice, in considering such a task, is that Sue's winning the community lottery fifty years running would be a circumstance that would immediately rouse our *suspicions*. More precisely, it would be a circumstance that would motivate us to find some *special explanation*—the lottery is rigged or biased or a guardian angel is smiling on Sue or some such. We can easily imagine ourselves exclaiming, 'Sue has won the lottery every year for the past fifty?! There has to be some explanation!' What this shows is that we are not inclined to treat Sue's winning the community lottery every year for the next fifty as being on an explanatory par with all the other possible outcomes. Furthermore, if Sue really did win the community lottery fifty years running, then it might be very *likely* that there is some special explanation for this—that the lottery is rigged in her favour, say. Indeed, the probability of this occurring by chance, rather than design, may be so low that we tend to neglect it altogether. If we do neglect this possibility and treat Sue's winning the community lottery fifty years running as a circumstance that would demand special explanation then, given my account of justification, we may regard ourselves as having justification for believing that this outcome won't obtain.

Put slightly differently, if Sue won the community lottery every year for the next fifty, then the probability, given my evidence, that there would be some special explanation for this may be very high—it may be 'for all intents and purposes' *certain* that some special explanation could be found. We might permit ourselves to say, in a case like this, that the proposition that Sue won't win the community lottery every year for the next fifty is 'for all intents and purposes' normically supported. But, when all is said and done, 'for all intents and purposes' normic support is *not* the same thing as normic support, and Sue's winning the community lottery every year for the next fifty *is* something that could occur purely by chance. There is going to be some sequence of winning tickets for the next fifty community lotteries and it might just as well be this sequence as any other.

When it comes to the mammoth lottery, of course, the circumstance in which Sue's ticket wins will not, in and of itself, arouse our suspicions, or prompt us to go looking for special explanations. What sets apart Sue's winning the next fifty community lotteries and Sue winning the mammoth lottery is that the former outcome exhibits a salient *pattern*, while the latter outcome does not. And there is considerable evidence to suggest that people are strongly disinclined to accept that event sequences exhibiting a recognisable pattern could arise through a purely random process, expecting such processes to produce outcomes that are unpatterned. People tend, in general, to perceive patterned sequences as non-chance and as demanding special explanation.

One suggestive study, performed by Falk and Konold (1997), demonstrated that the average confidence with which subjects judged sequences of binary values to be randomly generated was strongly correlated with the average *time* it took for subjects to successfully memorise such sequences, and that both judgements of randomness and memorisation performance were well predicted by a measure of encoding difficulty. It is often noted anecdotally that people have a general tendency to disregard the possibility of 'coincidental' events occurring purely by chance—preferring, in many cases, even conspiratorial or supernatural explanations over such a concession. Some further research tends to corroborate this more general impression.[9]

It is often observed, relatedly, that people exhibit a bias against the possibility of patterned outcomes in actual lotteries. People tend to dismiss the possibility that this week's lottery numbers will be '123456', but are open to the possibility that the numbers might be, say, '497235'—even though either sequence is equally likely to come up. In my view, this may be the very same bias that drives our intuitions about iterated lottery cases. Indeed, I suspect that a puzzle very like Vogel's could be set up simply by contrasting patterned and non-patterned outcomes within a single lottery. It will seem intuitive to many that I can legitimately assert and act upon the proposition that this week's lottery numbers won't be '123456', but not the proposition that this week's lottery numbers won't be '497235'.

We don't, of course, have any special insight into the outcomes of iterated lotteries that we lack when it comes to individual lotteries. On my view, we don't have justification for believing anything about the outcomes of a sequence of fair iterated lotteries, and any inclination to think that we can justifiably dismiss certain outcomes—such as Sue winning every one—is simply a manifestation of a more general human bias against the possibility of patterned outcomes arising through a random processes.[10]

It's worth noting finally, that it does seem possible (as this story would seem to predict) to divest oneself of the troublesome intuitions about iterated lotteries. Indeed, the comparison between the iterated lottery and the mammoth lottery can, no doubt, be helpful here. Once I do genuinely conceptualise Sue's winning the community lottery fifty years

[9] There is some evidence indicating that people tend to perform badly when asked to generate random sequences or to predict the outcomes of random processes—in particular, people appear to exhibit a bias against sequences with a large amount of repetition and tend to overestimate the frequency of alternation (see, for instance, Ross and Levy 1958). People's *judgements* regarding the randomness of presented sequences have also been the focus of research (Falk 1981, Lopes and Oden 1987 , Falk and Konold 1997). Findings tend to corroborate the existence of a general bias to associate randomness with excessive alternation. Naturally, the interpretation of the experimental findings is delicate and has been subject to dispute—for overviews see Bar-Hillel and Wagenaar 1991, Nickerson 2002.

[10] Though, as noted above, this bias is not wholly irrational—it may well be true that patterned outcomes are in general more *likely* to have special non-chance explanations than unpatterned ones.

running as just one amongst 10^{100} possible winning ticket combinations, each equally probable and equally plausible, then I will lose any tendency to think that I can simply rule this out in advance. And one can train oneself to overcome the more general bias against the possibility of random processes generating patterned outcomes. The mathematician Marcus du Sautoy once advised people to select consecutive numbers when playing the lottery on the grounds that others won't do this and, as such, if the numbers were to come up, you'd be less likely to have to share the winnings.[11] (Though whether this really is sage advice is unclear to me, given that, if you really *did* win with a series of consecutive numbers, almost everyone would suspect foul play!)

Part of my aim here has been to cast some well-known epistemic puzzles in a new light. My primary aim, though, has been to use these puzzles as a way of pressuring or testing some of the ideas developed in Chapter 2—in particular, the claim that normic support is necessary for justification. As should be evident, the claim fared better with respect to some puzzles than others. It offers a straightforward resolution of the lottery paradox—and a resolution that is consistent with multiple premise closure and does not risk any immediate slide into scepticism. When it comes to Hawthorne's lottery-driven scepticism, however, what it offers is less definitive, and may prove too concessive, and too precarious, for some tastes. The claim does offer a treatment of Harman's lottery puzzle and one that is superior, or so I have argued, to the main extant alternative exploiting the notion of sensitivity. Finally, while the claim is not consistent with all of the intuitions at play in Vogel's puzzle, it offers a very plausible story about what is driving the intuitions with which it clashes and, as I've argued, this is the best that we can hope for when it comes to this puzzle.

Philosophers will, of course, disagree on the relative importance of these different puzzles. Some may be inclined to dismiss one or more of them altogether. But I'm not concerned here to extract any kind of final verdict about my account and its capacity to solve epistemic puzzles. I will leave things here with the above 'scorecard' and allow these sorts of questions to remain open.

[11] <http://www.dailymail.co.uk/news/article-2052688/Why-picking-consecutive-numbers-increases-chance-Lottery-jackpot.html>.

4

Multiple Premise Closure

4.1. The Preface Paradox

The claim that normic support is necessary for justification resolves the lottery paradox while allowing us to retain the principle of multiple premise closure—the principle according to which, if I have justification for believing each of a series of propositions, then I also have justification for believing their joint deductive consequences. Just four years after Kyburg first proposed the lottery paradox, another powerful threat to the principle of multiple premise closure emerged in the form of the *preface paradox*, devised by David Makinson (1965). Imagine an author who completes a long, ambitious book on some topic, but modestly concedes, in the preface, that there are doubtless some false claims within. Suppose the author was very assiduous in researching his book—so much so that every individual claim in the book is justified by the evidence he has managed to unearth. Nevertheless, given the sheer scope and size of the book, it is exceedingly unlikely that *all* of these claims will turn out to be true—and, as such, it seems that the author's concession in the preface is also perfectly justified. If multiple premise closure holds, then it looks as though the author must have justification for believing an out and out contradiction—namely, that all of the claims in the book are true and that some claim in the book is false.

Some of the subsequent literature on this paradox has fixated on features of Makinson's example that are, to my mind, incidental. Weatherson (2005) and Leitgeb (2014a), for instance, attempt to engage with the paradox by suggesting that authors typically don't *believe* all of the claims made in scholarly works—rather, they put them forward as 'best guesses' or merely believe that some high proportion of them are true. Ryan (1991) suggests that the situation in which an author has evidence that justifies every single claim in a long ambitious book is very unrealistic and, as such, we should be suspicious of our own judgements about the case. Whether or not these suggestions are right, the paradox has nothing essential to do with *books* per se. It will arise whenever one's evidence justifies each of a set of propositions, and makes each one likely to be true, while also suggesting that they are unlikely to all be true—and this kind of set up is not at all unusual.

Suppose I'm holding a large dinner party to which I've invited 100 guests. Suppose that all 100 have replied saying that they will attend. Suppose, finally, that I know all of the invited guests to be honest and trustworthy and well-meaning and I have no reason whatsoever to suspect, of any one of them, that he or she won't be there. Presumably, then, I have justification for believing that Chuck will attend, that Barbara will attend, that Rosemary will attend, and so on for all the others. If multiple premise closure holds then I must also have justification for believing that *all 100* invited guests will attend. But surely this *wouldn't* be a sensible thing to believe. Surely, if I'm being realistic, I should believe the exact opposite—there will be fewer than 100 people at the dinner party (this example is based on one given by Hawthorne 2003: 48–9).

We can make this impression more vivid by attaching some probabilities to the propositions in question. Let's say that, given all the positive replies and the general reliability and trustworthiness of the people involved, it is 95% likely that Chuck will attend, 95% likely that Barbara will attend, and so on. If we suppose, for simplicity, that each person's attendance is independent of the others—so whether one person attends has no bearing on whether any other person attends—then the probability that all 100 invited guests will attend is simply equal to the product of the 100 individual probabilities—$0.95^{100} \approx 0.006$. The probability that *not* all 100 invited guests will attend is equal, then, to $1 - 0.95^{100} \approx 0.994$.

Surely, given these figures, I have justification for believing that not all 100 invited guests will attend. If this still fails to convince, we could embellish the example as follows: imagine I've organised countless such dinner parties in the past and, on every previous occasion, without fail, some invitees who responded positively to their invitations have failed to show. I have no reason to think that my latest dinner party will be any different. Surely it would be reckless of me to ignore this powerful inductive evidence to the effect that not everyone will attend.

Multiple premise closure has landed us with the conclusion that I have justification for believing that all 100 invited guests will attend. The probabilities and the background inductive evidence strongly suggest that I have justification for believing that not all 100 invited guests will attend. By a further application of multiple premise closure, it follows that I have justification for believing an out and out contradiction—namely, that all 100 invited guests will attend and not all 100 invited guests will attend. Since I clearly don't have justification for believing this, our only option is to reject the principle of multiple premise closure.

This is, in effect, a preface-type paradox set out in what I find to be a particularly persuasive way—a way that owes much to the compelling presentation of the paradox in Christensen (2004: chap. 3). And the reasoning is, I think, *very* persuasive—the most persuasive argument against multiple premise closure of which I'm aware. If we accept the risk minimisation conception of justification, then we should endorse every step of this reasoning as it stands and conclude that multiple premise closure is mistaken. If, however, we take seriously the idea that normic support is necessary for justification, then there are a few steps that we might find questionable.

If I have justification for believing that Chuck will attend, that Barbara will attend, etc. and normic support is necessary for justification, then each of these propositions must be normically supported by my evidence. And this would indeed appear to be so. If Chuck failed to attend, in spite of his positive reply, then there would have to be some explanation for this—he was ill, there was a family emergency, he got into a car accident, etc. The same goes, of course, for Barbara and for Rosemary and for each of the other invited guests. Given the model of normic support developed in section 2.4, normic support is closed under multiple premise deductive consequence, in which case the proposition that all 100 invited guests will attend must also be normically supported by my evidence. And a moment's reflection should confirm that the proposition is indeed normically supported. If the proposition turns out to be false, then there must be *some* guest who fails to attend and, since every guest gave a positive reply, there would have to be some explanation as to what happened—illness, family emergency, car accident, etc. The low evidential probability of this proposition owes, in a sense, to the myriad different ways in which it could turn out to be false. Its normic support owes to the fact that each of these myriad ways is one that demands explanation. There is no incoherence here.

What, then, of the proposition that *not* all 100 invited guests will attend? This proposition, as we know, is very likely to be true but, once we've given up the risk minimisation conception, nothing immediately follows about its justificatory status. In fact, if normic support is necessary for justification, then we should reject the supposition that this proposition is justified—it is not normically supported by any evidence that I possess. If it turned out that all 100 guests really did attend the party then, given that they're all honest and reliable and said they were coming, this would not be an outcome requiring any special explanation. This is simply an example of an unlikely event that would not be in any way abnormal.[1]

What about the embellished case in which I also have strong inductive evidence to the effect that not every guest will attend? Suppose, once again, that I've held countless dinner parties before and every time some invited guests who said they were coming failed to show up. And suppose that, on this occasion, to my utter amazement, everyone does show. There is *some* temptation to think that a special explanation might be needed here. I could certainly imagine searching for an explanation at this point—quizzing the guests as to whether they were put under some special pressure to attend etc.—and I could well find one. At the end of the day, though, there *needn't* be any explanation out there to be found. Full attendance is just one of the 2^{100} possible attendance patterns that could have eventuated at the dinner party—and this might just as well have turned out to be the actual pattern as any other (in fact this is, by some way, the most likely of the 2^{100} patterns!). The background evidence might prompt me to go searching for an explanation, but it doesn't generate a need for one. Whatever the overall effect of this evidence is, it doesn't normically support the proposition that some invited guest will fail to attend.

[1] The principle that unlikely things *never happen* is sometimes attributed to Cournot (1843). A slightly softened version of this principle might be that it's always exceptional or abnormal for unlikely things to happen. In a way, I think such sentiments are the exact *opposite* of the truth—almost *everything* that happens is, suitably described, an unlikely thing, and unlikely things are almost *never* exceptional or abnormal.

It may be likely that I'll stand up some time in the next hour. But when I do eventually stand, it will be at some particular instant, specifiable in terms of hours, minutes, seconds, milliseconds, and more precisely still. And the chance of my standing at this particular instant will, of course, have always been very low. In a way, then, no matter when I get up, something exceedingly unlikely will have happened—but nothing abnormal or exceptional will. On the risk minimisation conception of justification we have a kind of standing entitlement for believing, of any unlikely thing, that it won't occur. In a way, it is extraordinary to suppose that we have any such entitlement.

Furthermore, the formal model of normic support places certain restrictions on how it can be distributed in a case such as this. Given this model, no body of evidence can be constituted in such a way as to normically support the conclusion that some invited guest will fail to attend while simultaneously normically supporting the attendance of each invited guest—no body of evidence can be constituted in such a way as to provide simultaneous normic support for each of an inconsistent set of propositions. Either the background inductive evidence is so strong as to undermine the normic support for the attendance of at least some particular invited guests, or it is not strong enough to normically support the proposition that some invited guests will not attend.[2]

In one way, the lottery paradox and the preface paradox are just opposite sides of the same coin. Both paradoxes concern a large set of propositions that are individually evidentially likely. In the lottery paradox it is effectively stipulated that I have justification for believing that some member of the set is false. It is then suggested that I also have justification for believing, of each member in the set, that it is true, in virtue of the fact that this is highly evidentially likely. From the suggestion and the stipulation it follows that multiple premise closure must be abandoned. In the preface paradox it is effectively stipulated that I have justification for believing, of each member of the set, that it is true. It is then suggested

[2] What if we were to imagine that there were 500 or 1,000 invited guests or even more? The more invited guests there are, the more astonishing full attendance would seem and the more uncomfortable it becomes to insist that this would require no special explanation. As long as the attendance of each individual guest is normically supported, the formal model will of course churn out the result that less than full attendance is not normically supported, irrespective of how many guests are invited. But this doesn't quite settle things—we might now worry that the formalism is not being faithful to the informal characterisation of normic support in terms of what does and what does not call for explanation. Worries of this kind are quite legitimate—but the cases under consideration should, I think, be ultimately understood in a way that is consistent with what the formalism predicts.

Suppose I have invited n guests to a dinner party, they have all replied saying that they will attend, and they do in fact all show up. What we would expect is that, as n increases, so too will the *probability*, given our evidence, of there being some special explanation for the full attendance—a special explanation may be rather unlikely in the case of 5 guests, somewhat more likely in the case of 50, and very likely in the case of 500. This probability will never reach 1 however, and, thus, will never reach a point at which a special explanation is *required* per se. What can happen however is that the probability of there being no special explanation may become so vanishingly small that 'for all intents and purposes' an explanation is required. In this case, we might even say that less than full attendance will be 'for all intents and purposes' normically supported by my evidence, to use the turn of phrase introduced in section 3.4. But, as noted there, normic support 'for all intents and purposes' is *not* normic support and, for the purposes of theorising, the two must be clearly separated.

that I also have justification for believing that some member of the set is false, in virtue of the fact that this is highly evidentially likely. From the suggestion and the stipulation it follows, once again, that multiple premise closure must be abandoned. In the two paradoxes the roles of the suggestion and the stipulation are effectively swapped around. There is no way of refuting multiple premise closure *just* with stipulations—a suggestion is needed in each case. My theory of justification has us reject both of the suggestions—and, thus, appears to offer a way of safeguarding multiple premise closure in the face of both paradoxes.

But there is a certain asymmetry between these paradoxes as well. If we maintain multiple premise closure in the face of the lottery paradox, we have to concede that a high evidential probability does not guarantee justification. If we maintain multiple premise closure in the face of the preface paradox, we have to concede this *and more*—namely that a low evidential probability does not guarantee a *lack* of justification. Given multiple premise closure, I must have justification for believing that all 100 invited guests will attend my dinner party. I may deny that I also have justification for believing its negation and thereby avoid the full paradoxical conclusion—but there's no escaping the fact that I have already accepted something that seems deeply troubling.

As I noted in section 2.4, even if we take normic support to be necessary for justification and take normic support to be closed under multiple premise deductive consequence, it won't follow that justification is closed under multiple premise deductive consequence. The claim that normic support is necessary for justification is compatible with there being other requirements on justification that are not closed under multiple premise deductive consequence—in particular, it is compatible with there being an evidential probability requirement on justification. Call the following the *hybrid* theory of justification: one has justification for believing P iff (i) P is normically supported by one's evidence and (ii) the probability of P, given one's evidence, exceeds a threshold t. On this theory, justification requires two logically independent components to fall into place. According to (i), in order for one to count as having justification for believing something, any error would have to be an *explicable* error. According to (ii), the *chances* of (explicable) errors must, in addition, be suitably low.[3]

[3] A hybrid theory of this kind is described by Peijnenburg and Atkinson (2014: s. 3). Peijnenburg and Atkinson consider this theory as part of a discussion of a phenomenon

In a sense, the hybrid theory *overdetermines* the resolution of the preface paradox. The theory predicts that I lack justification for believing that all 100 invited guests will attend my dinner party, since this is too unlikely given my evidence. It also predicts that I lack justification for believing its negation—that some invited guests will not attend my dinner party, since this is not normically supported by my evidence. The theory forces us to give up on multiple premise closure. I may be justified in believing, for each individual guest, that he or she will attend—these propositions are both normically supported and, individually, run only a small risk of error. But, as we begin to conjoin these propositions, the error risk begins to accumulate until it reaches a point at which justification is precluded. Just how many of these propositions can be conjoined before justification gives out will depend upon where we set the threshold value t. (If we stick with the assumptions that these propositions are independent and each has an evidential probability of 0.95 then the maximum number that can be justifiably conjoined will be the greatest integer n such that $n < \log_{0.95}(t)$).

If normic support were both necessary and sufficient for justification, then the closure of justification under multiple premise deductive consequence would be assured. Call this the *normic* theory of justification: one has justification for believing P iff P is normically supported by one's evidence. Justification, on this theory, becomes a kind of 'no-risk', rather than a 'low-risk' notion. This is not to say, of course, that justification carries a complete guarantee against error. Rather, it carries a guarantee against error of a certain kind—error that is not explicable in terms of extenuating circumstances.

The normic theory 'resolves' the preface paradox simply by denying that I have justification for believing that some invited guests will fail to attend. The theory concedes that I have justification for believing that all 100 invited guests will attend, in spite of its vanishingly small evidential

they term the 'diminishing influence of grounds': roughly, the probability of the final proposition in a chain of probabilistic support, where each proposition probabilistically supports its successor, becomes less and less dependent on the probability of the first proposition as the chain lengthens. On the risk minimisation theory of justification, then, the degree of justification of the final proposition in a chain of justificatory support, becomes less and less dependent on the degree of justification of the first proposition as the chain lengthens. Whether justification still exhibits this kind of property on the hybrid theory will depend on how we try to capture the notion of *justificatory strength* within the theory. In fact, as I shall discuss in Chapter 5, it is far from clear how this should be done.

probability. In believing that Chuck will attend, that Barbara will attend, etc., I run no risk of a brute or inexplicable error. Thus, as these claims are conjoined, there is no risk of the relevant kind to accumulate. My final conclusion that all 100 invited guests will attend is still perfectly free from error risk of this kind.

As I noted in Chapter 2, believing or asserting that a proposition P is *likely* often functions as a kind of fallback commitment that one can make when one is unwilling to believe or assert P itself. The hybrid theory can offer a straightforward account of this practice. When one has justification for believing P it would typically be the case, on the hybrid theory, that one also has justification for believing that P is likely, given one's evidence. The converse, though, will not hold—just because one has justification for believing that P is likely, given one's evidence, it won't follow that one has justification for believing P. On the normic theory, however, believing or asserting that P is likely won't always be a softer option than believing or asserting P itself, though it may function this way on occasion. On the normic theory, there will be clear cases in which one has justification for believing P, but does not have justification for believing that P is likely given one's evidence—and preface paradox cases will be amongst them.[4]

The preface paradox leaves us with two options: first, we can reject multiple premise closure for justification, as the risk minimisation and hybrid theories would have us do. Second, we can accept that one may have justification for believing propositions that are exceedingly evidentially unlikely, as the normic theory would have us do. The former will, no doubt, strike many as the better deal. Many have grown accustomed

[4] To assert, in a single breath, that P is true and that P is not likely will sound odd or conflicted. Perhaps this gives us reason to reject the normic theory—to hold that high evidential probability really must be necessary for justification. On closer inspection, though, it's unclear whether such a lesson should be drawn. For it also sounds odd or conflicted to assert, in a single breath, that P is true and that P is not *certain* (see for instance Williamson 2000: 254, Stanley 2008: s. 4). Not often, though, is this taken as a reason to reject *fallibilism*—to hold that evidential certainty is necessary for justification.

Further, an unqualified assertion of P clashes with even weaker concessions than this—such as an assertion that P is not certain to be certain, or not certain to be certain to be certain, etc. The fact that it sounds strange to assert both P and that P is not likely is an instance, then, of a broader phenomenon, the general significance of which is very far from obvious. One thing that it does show, perhaps, is how easy it is to put oneself into a frame of mind (however fleetingly) in which outright assertions seem inappropriate.

to giving up on multiple premise closure, and the idea that one could have justification for believing something that is evidentially unlikely seems unthinkable, in a way. In the remainder of this chapter I will argue that the former option is not as attractive as it might seem at first and that the latter is not as unattractive. I won't attempt, though, to reach any definitive conclusion and, thus, won't reach any final view as to whether the hybrid or normic theory is to be preferred.

4.2. The Cost of Denying Closure

If multiple premise closure fails, then there will be situations in which one is epistemically permitted to believe each of the premises of a deductively valid argument, and to accept its validity, and yet epistemically obliged to remain agnostic about its conclusion. But to self-consciously adopt such a combination of attitudes would seem very strange. Suppose you are leafing through my dinner party invitation list and we engage in the following conversation:

YOU: 'Is Chuck coming?'
ME: 'Yes.'
YOU: 'And Barbara?'
ME: 'Yes. Her too.'
YOU: 'And Rosemary?'
ME: 'Yes.'

and so on for the remaining ninety-seven people and then . . .

YOU: 'And those are all the invited guests?'
ME: 'Yes.'
YOU: 'So it follows from what you've said that everyone is coming?'
ME: 'Of course that follows.'
YOU: 'Then everyone is coming!'
ME: 'Oh no, I'm not saying that.'
YOU: 'But you just said, of each invited guest . . .'
ME: 'Yes, I know. I still stand by everything that I've said.'

If I truly believed, of each guest, that he or she will attend but refused to believe that every guest will attend, then these would be the sincere responses to each of your questions. And yet this pattern of responses seems exceedingly odd, suggesting that there is something amiss with the underlying pattern of attitudes.[5]

If I assert, of every guest on the list, that he or she will attend, then I *should* accept that every guest will attend—indeed, it seems that I've *already* committed myself to this and can't shirk the commitment by baulking at the conjunctive claim. It's almost as though, by committing to the attendance of each guest in turn, I am outlining, piece by piece, a certain picture of what the future holds in store—and the picture that emerges is one in which all invited guests attend. The final question is not, then, a request for some still missing piece of that picture—it seems, at best, a request for confirmation of what has already been said. This is why it seems so perplexing that I would refuse to give this confirmation, without then going back and retracting anything that I've said.

There are, of course, circumstances in which I would be anxious about accepting that all the guests will attend—when the possibility of car accidents or illness is playing on my mind or even, perhaps, when the low evidential probability of the proposition is salient to me. In such circumstances, though, I would also be reluctant to flat-out commit to the attendance of any individual guest. Indeed it would be natural, in such a state of mind, to resort to probabilistic qualification—to withdraw to claims such as 'Chuck will most probably attend', 'Barbara will very likely attend', etc.

These kinds of considerations are well known, and many find them persuasive. But imagining these kinds of scenarios and conversations is not the only way of agitating for multiple premise closure. In the remainder of this section, I will focus on a rather different kind of argument—one that is much less direct. It's very natural to think that, if my evidence justifies me in believing a proposition, then I should be free to use that proposition in my reasoning—use it in justifying further propositions which then also become available for use and so on. And

[5] It may be possible to give some alternative explanation as to why this exchange sounds as odd as it does. One might argue that my 100 assertions, confirming the attendance of each guest, serve to progressively narrow down the conversational 'common ground', modelled as a set of possible worlds (Stalnaker 1978). When I then reject your assertion that every guest will attend, the reason this sounds odd is that the proposition, by this stage in the conversation, is 'presupposed', being true at all worlds in the common ground. This explanation seems quite coherent and does not explicitly appeal to the principle of multiple premise closure as defined. It remains unclear, though, whether the denial of multiple premise closure could ever be reconciled with this kind of picture of the dynamics of conversation.

yet, the legitimacy of this sort of practice would seem to depend upon the following formal principle:

> If E provides justification for believing P and $E \wedge P$ provides justification for believing Q then E provides justification for believing Q.

If this principle fails, then my evidence could provide justification for believing P, and P, along with my evidence, could provide justification for believing Q while my evidence fails to provide justification for believing Q. Principles of this form are sometimes termed *cumulative transitivity* principles.

Consider, in addition, the following formal principle:

> If $E \wedge P$ provides justification for believing Q, then $E \wedge P$ provides justification for believing $P \wedge Q$.

We might call this *strengthening*. According to this principle, a proposition that is a *part* of a body of evidence can be freely conjoined to any proposition for which that evidence provides justification. Once we start to think in terms of potential formal principles for justification, these two principles can seem like plausible candidates. And yet, when it comes to the dinner party case under consideration, these principles can in fact be used to derive the very instance of multiple premise closure that drives the paradox.

Let I be the totality of my relevant evidence—the proposition that all 100 invited guests have replied to their invitations, indicating that they will attend the dinner party. Let A_1 be the proposition that guest #1 will attend, A_2 be the proposition that guest #2 will attend, and so on. We have it that I provides justification for believing A_1 and that I provides justification for believing A_2. We assumed above that each person's attendance is independent of the others—so whether one person attends has no bearing either way on whether any other person does. Given this assumption $I \wedge A_1$ will continue to provide justification for believing A_2. By strengthening it follows that $I \wedge A_1$ provides justification for believing $A_1 \wedge A_2$. But if I provides justification for believing A_1 and $I \wedge A_1$ provides justification for believing $A_1 \wedge A_2$ then, by cumulative transitivity, I provides justification for believing $A_1 \wedge A_2$.

The reasoning can be repeated. Given independence, $I \wedge A_1 \wedge A_2$ provides justification for believing A_3. By strengthening it follows that $I \wedge A_1 \wedge A_2$ provides justification for believing $A_1 \wedge A_2 \wedge A_3$. But if

I provides justification for believing $A_1 \wedge A_2$ and $I \wedge A_1 \wedge A_2$ provides justification for believing $A_1 \wedge A_2 \wedge A_3$ then, by cumulative transitivity, I provides justification for believing $A_1 \wedge A_2 \wedge A_3$—and so on. After 100 iterations, we arrive at the result that I provides justification for believing $A_1 \wedge \ldots \wedge A_{100}$ as required. Anyone who rejects multiple premise closure in order to resolve the preface paradox is obliged, then, to reject either cumulative transitivity or strengthening.

These two principles are not sufficient, in and of themselves, for us to derive multiple premise closure in full generality. For that, a further principle is needed:

> If E provides justification for believing P and E provides justification for believing Q then $E \wedge Q$ provides justification for believing P.

Justification is, of course, *defeasible*—just because a certain body of evidence provides justification for believing a proposition P, it doesn't follow that an enriched body of evidence will continue to do so. If E provides justification for believing P and $E \wedge D$ does not provide justification for believing P, then D might be said to *defeat* the justificatory relation between E and P. What the above formal principle does is to place a certain formal constraint upon which propositions can act as defeaters. According to this principle, if E provides justification for believing that P is true, vulnerable to defeat by D, then E cannot *simultaneously* provide justification for believing that D is true. Evidence, we might say, cannot be self-defeating. Principles of this form are sometimes termed *cautious monotonicity* principles.

If justification satisfies cumulative transitivity, strengthening, cautious monotonicity, and single premise closure, then it satisfies multiple premise closure. Suppose that E provides justification for believing P, E provides justification for believing Q, and that P and Q together entail R. By cautious monotonicity, $E \wedge Q$ provides justification for believing P. By strengthening, $E \wedge Q$ provides justification for believing $P \wedge Q$. If E provides justification for believing Q and $E \wedge Q$ provides justification for believing $P \wedge Q$ then, by cumulative transitivity, E provides justification for believing $P \wedge Q$. If $P \wedge Q$ entails R then, by single premise closure, E provides justification for believing R. If one's evidence provides justification for believing P and provides justification for believing Q and P and Q entail R, then one's evidence provides justification for believing R. This reasoning can easily be generalised for any number of premises. Anyone who rejects multiple premise closure, for any reason, is obliged to reject one or more of these principles.

Rather than appealing to imagined scenarios or conversations, the arguments I've offered here appeal only to formal principles from which multiple premise closure, or certain key instances thereof, can be derived. These arguments are not decisive. One who rejected multiple premise closure is at liberty to reject these principles as well. But what the arguments do serve to illustrate, at the very least, is that the rejection of multiple premise closure can have downstream formal consequences that are by no means obvious from the outset. For many philosophers, resolving something like the preface paradox is, in the end, a matter of conducting a *cost–benefit analysis*—of carefully weighing up the advantages and disadvantages of each of the solutions on offer. While I'm sympathetic to this kind of approach, it would be a mistake to assume that all of the alternative solutions wear their costs on their sleeves. The costs of abandoning something like multiple premise closure are not yet well understood.

4.3. Risk Maximisation

If we hold on to the principle of multiple premise closure, then the preface paradox forces us to concede that we can have justification for believing propositions that run a high risk of falsity. In the dinner party example, I have justification for believing the proposition that Chuck will attend, the proposition that Barbara will attend, etc., and, given my evidence, each of these propositions runs only a low falsity risk. As these propositions are conjoined, multiple premise closure guarantees that my justification is preserved, while the falsity risk begins to accrue. Ultimately, we reach the result that I have justification for believing that all 100 invited guests will attend, even though the risk that this proposition is false completely overwhelms the chance that it is true.

One immediate question we might ask about this is whether it is even *possible* to believe a proposition while, at the same time, estimating its probability to be low. We are accustomed to thinking of probability estimates as indicating one's *degree of belief* in a proposition, which suggests that this shouldn't be possible.[6] But *believing* something with a

[6] Mark Kaplan (1995, 2013) has suggested that one can believe a proposition P while estimating its probability to be low. Kaplan works, however, within a broadly Bayesian

certain confidence and estimating it to have a certain *probability* are not at all the same. If we put aside our preconceptions, there is nothing at all incoherent about an utterance such as 'It may be unlikely that all the invited guests will come, but I believe that they will' or even 'It may be unlikely that all the invited guests will come, but I strongly believe that they will.' If I were busily setting 100 places at the dinner table and someone snidely remarked that I shouldn't bother with so many, I could easily respond in this way and there would be no reason not to take the response at face value.

While this may then be a possible attitude for me to have, one might still deny that it could ever be a *justified* attitude. 'It may be unlikely that all the invited guests will come but I'm justified in believing that they will' does, admittedly, sound more jarring. We are accustomed to thinking that the significance of a body of evidence is exhausted by the evidential probabilities that it imposes on propositions. As such, if a proposition P is unlikely, given one's evidence, then we naturally take this to imply that the evidence does not support P at all and supports its negation instead. If *that* were true then of course one couldn't be justified in believing P. But whatever is naturally 'implied' by the fact that P is unlikely, given one's evidence, it doesn't logically follow that one's evidence is stacked in some overall way against P—on the contrary, it is compatible with one's evidential situation being far more nuanced.

Another powerful source of resistance to the idea that we can have justification for believing propositions that are evidentially unlikely is the intimate link between justified belief and *rationally permissible action*. If I have justification for believing a proposition P then it should, in some circumstances at least, be rationally permissible for me to act on the assumption that P is true—to use P as a premise in my practical reasoning and choose actions that will lead to beneficial consequences if P is true. But surely it could only be reckless to act on an assumption that is likely, given my evidence, to be false.

framework in which subjective probabilities are equated with degrees of confidence or belief. As such, he goes on to make further claims that seem to me not to be correct—that one can believe P while being confident that P is false or while investing a low degree of belief in P. In my view, subjective probabilities reflect one kind of specialised judgement that we can choose to make about a proposition, but they are not a foundational aspect of our intellectual lives, and are as separate from degrees of confidence and degrees of belief as they are from outright belief. I will have more to say about degrees of belief in section 8.4.

The decision of how many places to set at the dinner table may be of little consequence, but the question of whether all 100 invited guests will attend could easily assume greater practical significance. Suppose my dinner party is to be catered. Should I tell the caterers to prepare for 100 people? What if they charge substantially more to cater for parties of 100 or larger? Should I *still* tell them to cater for 100 and just bear any additional cost? If I have justification for believing that all 100 guests will attend, there is pressure to think that this should be rationally permissible. But telling the caterers to prepare for 100 would seem to violate the norms of decision theory. By acting in this way, I would be running a very high risk of one adverse outcome—food and money being wasted—in order to avoid a very low risk of another adverse outcome—some guests going hungry. The numbers, we might think, just don't add up.

Decision theory was originally developed for a very circumscribed setting—to determine how one might place advantageous bets when playing certain games of chance. The theory is now taken by many to be all but universally applicable—an attitude that invites us to view all actions and decisions as 'bets' on what the world is like. But the number of people attending my dinner party is not the outcome of some game of chance—and to pretend that it is is to ignore some important aspects of the situation. Insofar as I trust my friends, I *must* tell the caterers to prepare for 100 people. If I tell them to prepare for fewer, then I am displaying mistrust—perhaps not directed at any particular person, but mistrust nonetheless.

Further, if I do tell the caterers to prepare for 100 then, while it may be almost certain that fewer than 100 people will show up, there is a sense in which I would not ultimately be to *blame* for the money and food that would be wasted in this event. The blame for this, it seems, should lie with those guests who failed to attend, despite giving assurance that they would. If, on the other hand, I tell the caterers to prepare for fewer than 100 people and, against the odds, all 100 people do show up then *I* would bear the blame for guests going hungry. This is an outcome that I have responsibility for preventing, while preventing the outcome in which money and food is wasted seems not to be something that I am fully responsible for—part of the responsibility for this rests with the individual guests. By telling the caterers to prepare for 100 I am ensuring that my own responsibilities are met.

These observations are not unrelated to the fact that my evidence normically supports the proposition that all 100 invited guests will attend. If this turns out to be false, then this will be due to some extenuating circumstance—a particular invited guest was ill or had a family emergency etc. If I do decide to cater for 100 and some money and food is wasted then this too will be due to an extenuating circumstance, at least in part. This outcome will not be solely attributable to my action, but will be the result of my action in combination with further factors outside my control.

I don't mean to suggest here that telling the caterers to prepare for 100 is the *recommended* course of action in this scenario—perhaps it would be better, all in all, if I didn't think about the decision in terms of a division of responsibilities, but regarded it as a kind of gamble instead. But whatever else we want to say about it, this action would, I maintain, be rationally permissible—and this is consistent with my having justification for believing that all 100 invited guests will attend.

What if the adverse consequence that I'm risking is something that is very severe—what if I would be bankrupted or sued if I told the caterers to prepare for 100 and fewer guests showed up? If I'm aware of these potential consequences then, clearly, I should just hold my tongue and not tell the caterers to prepare for 100. But this should not incline us to think that I lack justification for believing that all 100 invited guests will attend. After all, if a great deal was hanging in the balance I should not say, of even a *single* guest, that he or she will attend—but this doesn't incline us to think that I lack justification for believing that the guest will attend. If the stakes are high, it may not be rationally permissible to put even a justified belief into action. I will have more to say about the relation between justification, stakes, and actionability in Chapter 5.

Aside from these preface paradox-type cases of accumulating risk, there may be further cases in which we find ourselves inclined to believe propositions that turn out, on reflection, to be evidentially unlikely. Suppose, once again, that a bus causes some harm on a city street. Suppose that someone we regard as very reliable and trustworthy tells us that she saw the incident and that the bus involved was owned by the Blue-Bus company. It would be natural to take this eyewitness at her word and come to believe that the bus involved was a Blue-Bus bus. Suppose we then come across some statistical evidence to the effect that only 5% of the buses operating in the area, on the day in question, were Blue-Bus buses. Should we now begin to *doubt* whether the bus involved was a Blue-Bus bus—to doubt the witness's words? It seems to me that this would not necessarily be the right reaction—if we trusted the testimony before we received the statistical evidence, we could reasonably continue to trust it afterwards.

There are, of course, many things we could discover that might call a piece of testimony into question. We could, for instance, uncover evidence that provides justification for believing that the proposition being testified to is false. Alternatively, we could uncover evidence that provides justification for believing that the testifier is unreliable or untrustworthy. But the evidence that only 5% of the buses operating in the area on the day in question were Blue-Bus buses seems not to do either of these things. It doesn't provide justification for believing that the bus involved was not a Blue-Bus bus—it would be wrong to make such a judgement on the basis of this evidence alone. Neither does it provide justification for thinking that our eyewitness is untrustworthy or unreliable—to leap to such a conclusion would be exceedingly unfair. Surely, then, I should still be justified in trusting this testimony and in believing that the bus involved was a Blue-Bus bus.

On this way of thinking, learning that only 5% of the buses operating in the area on the day in question were Blue-Bus buses does not affect my justification for believing that the bus involved was a Blue-Bus bus. But it does, of course, affect the *evidential probability* of this proposition. Let T be the proposition that the witness testified that the bus involved was a Blue-Bus bus and let S comprise our statistical evidence about the proportion of Blue-Bus buses operating in the area on the day in question. Let B be the proposition that the bus involved was a Blue-Bus bus. If Pr is our evidential probability function, given background evidence, then it's plausible that $Pr(B \mid S) = 0.05$ and $Pr(\sim B \mid S) = 0.95$.

Finally, we need some measure of the statistical reliability of our eyewitness. Suppose that, given the proximity of the witness, the duration of the incident, the lighting conditions, etc., the evidential probability of the witness correctly identifying a Blue-Bus bus is 85%. That is, suppose that the evidential probability of the witness testifying that the bus was a Blue-Bus bus, given that it really was a Blue-Bus bus, is 85% and the evidential probability of the witness testifying that the bus was a Blue-Bus bus, given that it was not a Blue-Bus bus, is 15%. This supplies us with values for the so-called 'Bayesian likelihoods': $Pr(T \mid B \wedge S) = 0.85$ and $Pr(T \mid \sim B \wedge S) = 0.15$. The probability that the bus was a Blue-Bus bus, given our total evidence—$Pr(B \mid T \wedge S)$—can then be calculated using

Bayes' theorem: $\Pr(B \mid T \wedge S) = (\Pr(B \mid S).\Pr(T \mid B \wedge S))/((\Pr(B \mid S).\Pr(T \mid B \wedge S) + \Pr({\sim}B \mid S).\Pr(T \mid {\sim}B \wedge S)) = (0.05 \times 0.85)/((0.05 \times 0.85) + (0.95 \times 0.15)) \approx 0.23$. Perhaps what this shows is that one would not be justified in continuing to believe that the bus involved was a Blue-Bus bus. By neglecting the evidence that only 5% of the buses operating in the area on the day in question were Blue-Bus buses we might be accused of committing the *base rate fallacy*—something I will return to. But if we do concede that this is a case in which a proposition is justified, in spite of its low evidential probability, then the corresponding concession in preface paradox-type cases will seem much less costly.

In fact, the normic theory of justification will treat both kinds of case in the same way. As well as providing a certain amount of probabilistic support for the proposition that the bus involved was a Blue-Bus bus, the eyewitness testimony also provides normic support for this proposition. If the bus involved was not a Blue-Bus bus, in spite of the witness testimony, then there would have to be some explanation for this—the sun was in her eyes, she was momentarily distracted, something partially obscured her view, she was lying, etc. In all of the most normal worlds in which the witness testifies that the bus involved was a Blue-Bus bus, the bus involved was a Blue-Bus bus.

Normic support is of course *defeasible*—just because a body of evidence provides normic support for a proposition it does not follow that it will continue to do so when new evidence is added. Suppose we were to learn that some rival bus companies have painted their buses to closely resemble Blue-Bus buses. Given that the witness testified that the bus involved was a Blue-Bus bus *and* some rival bus companies have painted their buses to closely resemble Blue-Bus buses, there need be nothing left to be explained, in the event that the bus involved was not a Blue-Bus bus. In some of the most normal worlds in which the witness testifies that the bus involved was a Blue-Bus bus and some rival bus companies have painted their buses to closely resemble Blue-Bus buses, the bus involved was not a Blue-Bus bus, but one of these disguised rivals instead.

In contrast, the evidence that only 5% of the buses operating in the area on the day in question were Blue-Bus buses does not negate the need for an explanation if the bus turns out not to have been a Blue-Bus bus. On the contrary, an explanation of exactly the sorts considered above would still be needed. The statistical evidence might make it more probable that the bus involved was not a Blue-Bus bus and thus more probable that some explanation will be required—but this is not at all to remove the requirement for explanation. Once one is struck by the thought that there would have to be more to the story if the bus turned out not to be a Blue-Bus bus, then no information about the proportion of Blue-Bus buses operating in the area on the day in question would—or *should*—shake that thought.

There is also a more formal point to be made here. Observe that the fact that the witness testified that the bus involved was a Blue-Bus bus does not normically support any propositions about the proportion of Blue-Bus buses operating in the area on the day in question—it's consistent with any number of different views about that. But if the witness testimony does not normically support any propositions about the proportion of Blue-Bus buses operating in the area on the day in question, then propositions about the proportion of Blue-Bus buses operating in the area on the day in question are unable to defeat the normic support that the witness testimony supplies for the proposition that the bus was a Blue-Bus bus. More generally, if E normically supports P, and $E \wedge D$ does not normically support P, then E must normically support ~D. This, in any case, is easily proved given the possible worlds model of normic support that I've developed. This formal property is sometimes known as *rational monotonicity*—I'll have more to say about it in Chapter 7.

What if the proportion of Blue-Bus buses was even lower—what if only one in a thousand buses is a Blue-Bus bus or one in a million? Changing the numbers will not make any difference to the normic support that the testimony provides for the proposition that the bus involved was a Blue-Bus bus, but does it change our intuitions? I'm not claiming of course that one is *obliged* to trust the eyewitness and believe that the bus was a Blue-Bus bus. If one does give up this belief upon learning that the proportion of Blue-Bus buses is very low, then one need not commit any epistemic wrong. But holding on to this belief need not, I think, involve an epistemic wrong either. In a sense there is nothing at all *difficult to believe* about what the witness is testifying to. We know that a bus was involved in the incident and it might just as well have been the one Blue-Bus bus, as the witness says, as any one of the 9,999 (or more) other buses.

According to the normic theory, the eyewitness testimony provides justification for believing that the bus involved was a Blue-Bus bus, and this justification is not diminished or defeated by the information that only a small proportion of the buses operating in the area on the day in question were Blue-Bus buses. As mentioned above, some will urge that anyone who continued to believe that the bus involved was a Blue-Bus bus, even after discovering this statistical information, would be guilty of committing the *base rate fallacy* (see, for instance, Kahneman and Tversky 1972, Bar-Hillel 1980). The base rate fallacy is a very specific error of probabilistic reasoning—the error of neglecting base rate information when attempting to estimate how probable a hypothesis is made by a body of evidence. But by trusting the eyewitness to this incident, and believing that the bus involved was a Blue-Bus bus, I am not estimating the *probability* of anything. As we've seen, figuring out what to believe about a given situation is not at all the same as trying to figure out how to assign probabilities to propositions.

The present variant on the Blue-Bus case is, in fact, based upon an example introduced by Kahneman and Tversky in a classic early study on the base rate fallacy (see Kahneman and Tversky 1972). When asked to judge the probability of the proposition that is supported by the eyewitness testimony, subjects tend to offer estimates much higher than that supplied by the Bayesian calculation, suggesting a relative neglect of the statistical evidence (Kahneman and Tversky 1972, Lyon and Slovic 1976, Bar-Hillel 1980).

But whether subjects who respond in this way are genuinely guilty of committing a fallacy depends very much on how they are representing the problem to themselves. It may be more natural for subjects to think about the bearing of something like testimony in terms of normic, rather than probabilistic, support. And, when assessing the extent to which a hypothesis is normically supported by available evidence, *it is right* to neglect base rates—such information is genuinely irrelevant. The prompts typically used in these studies do ask for 'probabilities' or 'chances'—but this alone seems insufficient to control for this potentially interfering factor. We should not suppose that the meaning of the term 'probability' in ordinary usage invariably corresponds with what philosophers, psychologists, statisticians, etc. mean by it (Cohen 1977: §12)—particularly if we take seriously a plurality of evidential support relations. Some psychologists have recently taken the view that questions mentioning 'probabilities' or 'chances', particularly when they concern a single hypothesis or event, can be interpreted in a number of different ways by subjects (Cosmides and Tooby 1996: s. 1.4).

There are many and varied studies purporting to demonstrate a general human susceptibility to the base rate fallacy, though this conventional conclusion is sometimes challenged[7]—for critical overviews of this literature see Koehler (1996), Stanovich and West (2000), and Barbey and Sloman (2007). I don't mean to raise any general concerns about this work or the conclusion that is standardly drawn from it. It is interesting to note, though, that some of the best-known problems used in these studies share the potentially confounding feature I've identified here—namely that the target hypothesis is *normically*, as well as probabilistically, supported by the evidence supplied. This certainly seems to be so, for instance, in the diagnosis problem used by Hammerton (1973) and the light bulb problem used by Lyon and Slovic (1976). For other problems—such as the lawyer-engineer problem used by Kahneman and Tversky (1973) and the suicide and dream problems used by Bar-Hillel (1980)—it is much less clear whether this feature is present, though the temptation to try and reason normically may still be. I won't explore this further.

The preface paradox is one flashpoint at which the differences between the normic theory and the risk minimisation theory ignite. The risk minimisation theory, and the hybrid theory, would have us reject the principle of multiple premise closure and deny that I can have justification for believing that all 100 invited guests will attend my dinner party. The normic theory would have us retain multiple premise closure and grant that I can have justification for believing that all 100 invited guests will attend, its low evidential probability notwithstanding. I have made no claims here about which alternative carries greater benefits and fewer costs. What I have tried to argue is that the costs and benefits of these alternatives do not yet lie fully open to view. We are not in any position to 'weigh them up' once and for all.

[7] These who are sceptical of the base rate fallacy sometimes take issue with the Bayesian calculations that experimenters use to determine the 'correct' answers to the problems they set—such as the calculation used above. Some have suggested that these calculations themselves are somehow flawed (Jonakait 1983)—arguably putting themselves at odds with classical probability theory. Others have questioned the input values required to get such Bayesian calculations up and running, disputing in particular the prior probabilities that the experimenters generally assume to be appropriate (Levi 1981, 1983, Niiniluoto 1981). For a survey of some of this literature, see Koehler (1996) and Stanovich and West (2000). Unlike these commentators, I find no particular fault in the Bayesian calculation per se—I'm happy to accept that it supplies the correct answer *to a certain question*—but not the only question that might be asked about the bearing of the evidence provided (see also Cohen 1979).

5

Comparative Justification

In Chapter 4 I set out two theories. According to the normic theory of justification, one has justification for believing P iff P is normically supported by one's evidence. According to the hybrid theory of justification, one has justification for believing P iff P is normically supported by one's evidence and the probability of P, given one's evidence, exceeds a threshold t. As they stand, these theories supply truth conditions for *categorical* attributions of justification. But a full theory of justification must, ultimately, have something to say about justification *comparisons* as well. When it comes to comparisons, however, the hybrid theory faces a number of difficulties. While the normic theory has a relatively straightforward way of dealing with comparisons, one potential issue leads us to consider a significant variation on the theory.

5.1. Problems with Comparisons

As well as appraising the justification that one has for believing a single proposition, it is often natural to compare the justification that one has for believing distinct propositions—natural to say things like 'I have more justification for believing that Hillary and Norgay reached the summit of Everest in 1953 than I do for believing that Mallory and Irvine reached the summit of Everest in 1924', 'Bruce has more justification for believing that Sydney is the capital of New South Wales than he does for believing that Vaduz is the capital of Liechtenstein,' 'I have more justification for believing that there is water on Earth than I do for believing that there is water on Mars,' etc.

The risk minimisation conception, as set out in section 2.1, includes a straightforward account of justification comparisons: one has more justification for believing P than Q iff one's evidence makes P more probable than Q. But what is a hybrid theorist to say about such comparisons?

A natural first thought is perhaps the following: one has more justification for believing P than Q iff (i) one's evidence normically supports P more strongly than Q and (ii) one's evidence makes P more probable than Q. On this suggestion, justification comparisons require two components to align, much as categorical attributions of justification do. In order for a justification comparison to be true, a corresponding normic support comparison and evidential probability comparison must both be true. We might call this the hybrid theory of comparative justification.

As was demonstrated in Chapter 2, though, normic support and evidential probability comparisons can run in opposite directions—it's possible for a body of evidence E to normically support P more strongly than Q while making Q more likely than P. The laptop example has just this structure. If we hold the hybrid theory of justification comparisons, then we are obliged to say, in a case like this, that P and Q are *incomparable* with respect to justification—it's not the case that one has more justification for believing P than Q, or that one has more justification for believing Q than P or indeed that one has equal justification for believing both. On the hybrid theory of comparative justification, there is no guarantee that propositions can always be compared for their justifiedness. So it is, in general, when comparisons are *multi-parametric* and require two or more independent underlying comparisons to align.

The suggestion that propositions may be incomparable with respect to justification is not I think problematic per se. But there are certain cases, of course, in which it would be problematic for comparability to break down. Suppose that both P and Q have an evidential probability above the threshold t and that Q has a greater evidential probability than P. Suppose, further, that P is normically supported by my evidence and Q is not (for a suitable choice of threshold, the laptop example will once again provide an illustration of a case with exactly this kind of structure). The hybrid theory of justification predicts, in a case like this, that I have justification for believing P but not for believing Q—P meets the conditions for justification while Q does not. But the hybrid theory of comparative justification predicts that P and Q are *incomparable* with respect to justification. Surely, though, if I have justification for believing P, and I don't have justification for believing Q, then the two propositions *are* comparable with respect to justification—surely I have *more* justification for believing P than Q. This seems a non-negotiable logical connection between categorical and comparative attributions of justification.

Another possibility, perhaps, is to use evidential probability as the *sole* basis for justification comparisons and to treat normic support purely as a necessary condition on categorical justification attributions. The suggestion, more precisely, is this: one has more justification for believing P than Q iff P is more likely, given one's evidence, than Q. This is just thesis (b) of the risk minimisation conception—we might call this the risk minimisation theory of justification comparisons. Consider again a case in which Q is more likely, given my evidence, than P, P is more likely, given my evidence, than the threshold t, and P is normically supported by my evidence while Q is not. The hybrid theory of justification predicts that (i) I have justification for believing P and (ii) I do not have justification for believing Q, while the risk minimisation theory of justification comparisons predicts that (iii) I have *more* justification for believing Q than P. These predictions are clearly inconsistent.

What, then, if we were to use normic support comparisons as the sole basis for justification comparisons? What if one has more justification for believing P than Q iff one's evidence normically supports P more strongly than Q? This *normic* theory of justification comparisons will give the right results in the sort of case that we have just considered. But it, too, runs into logical problems when combined with the hybrid theory of categorical justification attributions. Consider now a case in which P is more likely, given my evidence, than Q. Suppose, further, that the justification threshold t lies in between the evidential probabilities of P and Q. Suppose, finally, that my evidence normically supports both P and Q but supports Q more strongly than P. The hybrid theory of justification predicts that (i) I have justification for believing P and (ii) I do not have justification for believing Q, while the normic theory of justification comparisons predicts that (iii) I have *more* justification for believing Q than P.

Perhaps one could develop some further, more complicated, account of justification comparisons to complement the hybrid theory—but the difficulties involved in doing this should not be underestimated. The two cases just described suffice to show that, if the hybrid theory of justification is correct, then neither evidential probability comparisons nor normic support comparisons can be necessary or sufficient for corresponding justification comparisons. The first case described above is one in which Q is more likely than P, but, assuming the hybrid theory of justification, I must have more justification for believing P than Q. Thus, an evidential probability comparison can be neither necessary nor sufficient for a corresponding justification comparison. The second case described above is one in which Q is normically supported more strongly than P, but, assuming the hybrid theory of justification, I must have more justification for believing P than Q. Thus, a normic support comparison can be neither necessary nor sufficient for a corresponding justification comparison.

It is, at the very least, difficult to see how the hybrid theory might be squared with an adequate account of comparative justification—and this, I think, is an important objection to the theory. It's worth noting, in passing, that this general kind of criticism may be widely applicable—and may even serve as a useful filter on prospective theories of categorical justification attributions. Many of the theories of justification that I've mentioned along the way, including the theories proposed by Sutton (2007), Bird (2007), Reynolds (2013), and Ichikawa (2014), would arguably be filtered out.

Sutton, Bird, Reynolds, and Ichikawa, as discussed in section 1.1, each seek to understand justification in terms of knowledge—for Sutton justification just is knowledge, for Bird and Ichikawa justification is a kind of 'would be' knowledge, and for Reynolds justification is the appearance of knowledge. But it's standardly thought that (propositional) knowledge does *not* permit of comparisons (see Stanley 2005: chap. 2). Given this, it's difficult to see how justification comparisons could be accommodated by any of these theories. Ichikawa does not mention comparative justification. Reynolds mentions it briefly—seemingly admitting that his account has no way of dealing with it (Reynolds 2013: 368). Sutton is more explicit—he states that, on his theory, justification comparisons are strictly speaking *nonsense* (Sutton 2007: chap. 2 n. 4). Sutton may be right to think that this is a consequence of his theory—but why this should not be regarded as a *reductio* is unclear (see McGlynn 2014: s. 3.1).

Unlike Ichikawa, Reynolds, and Sutton, Bird does make a genuine attempt to accommodate at least doxastic justification comparisons. Suppose a subject, with a certain set of mental states, comes to believe P. On Bird's theory, at a first pass, the belief will count as justified iff there is a possible world in which the subject has the very same mental states and comes to know P (Bird 2007: s. 2). For Bird, a belief is justified iff it counts as knowledge or is prevented from counting as knowledge by something non-mental. But if knowledge is an all-or-nothing status, then justification, so characterised, will also be an all-or-nothing status. What then could it possibly mean for one belief to be *more* justified than another?

What Bird suggests is that comparing the justification of beliefs is essentially a matter of comparing how close they come to satisfying the mental requirements for knowledge or of comparing how safely they meet those requirements. For any two beliefs that fail to meet Bird's condition for justification, one will count as more justified than the other just in case it comes *closer* to meeting this condition. For any two beliefs that meet Bird's condition, one will count as more justified than the other just in case it is *further* from violating the condition (Bird 2007: s. 9). In effect, when comparing beliefs that are unjustified, one will count as more justified than the other just in case it is closer to being justified and, when comparing beliefs that are justified, one will count as more justified than the other just in case it is further from being unjustified.

According to Bird, 'beliefs of all kinds (even those that constitute knowledge) may be graded for their degree of justification. But, since knowledge itself is a matter of hitting the bull's eye, it is an all-or-nothing affair and it is typically not appropriate to attach degrees to the concept of knowing' (Bird 2007: 108). On closer inspection, though, Bird's account does make it very puzzling why justification should permit of comparisons while knowledge does not. After all, we can easily compare beliefs for how close they are to knowledge or how far they are from ignorance. If corresponding judgements are *all* that underlie comparative attributions of justification, then why can we not also make comparative attributions of knowledge? Bird's theory, like others that attempt to understand justification in terms of knowledge, will ultimately struggle to explain this one striking asymmetry between the two notions—justification permits of comparisons, while knowledge does not. I won't pursue this criticism further here.

5.2. The Threshold Normic Theory

The normic theory of justification can be paired with a matching account of comparative justification—namely the *normic* theory of comparative justification that has already been mentioned: one has more justification for believing P than Q iff one's evidence normically supports P more strongly than Q. This combination will not run into the kinds of difficulties just discussed. Rather, the combination will ensure that, if one has justification for believing P but not for believing Q, then one has more justification for believing P than Q (and does not have more justification for believing Q than P).

One point, though, is worth making about this combined theory. Evidence E normically supports a proposition P more strongly than a proposition Q just in case the most normal worlds in which E is true and P is false are less normal than the most normal worlds in which E is true and Q is false. Clearly, though, if the most normal worlds in which E is true and P is false are *less normal* than other worlds in which E is true, then the most normal worlds in which E is true must be worlds in which P is true, in which case E normically supports P. That is, if E normically supports P more strongly than Q, it will follow that E normically supports P. On the normic theory of categorical and comparative justification, if one has more justification for believing P than Q, it follows that one has justification for believing P. When it comes to justification, any favourable comparison will commit one to a categorical attribution.

This prediction may seem surprising, for gradable adjectives do not, in general, have such a feature. If Roger is taller than Maria it certainly doesn't follow that Roger is tall—'Roger is taller than Maria but Roger is not tall' seems perfectly felicitous. With other gradable adjectives, though, the situation is perhaps less clear. If the sponge is *wetter* than the towel, for instance, then it does seem to follow that the sponge is *wet*—it sounds odd to say 'The sponge is wetter than the towel, but the sponge is not wet'. Consider also 'The hen on the left is more speckled than the hen on the right, but the hen on the left is not speckled', 'The red liquid is more poisonous than the pink liquid, but the red liquid is not poisonous'. It may be that 'justification' does belong in this category—it sounds somewhat odd, to my ear, to say 'One has more justification for believing P than Q, but one does not have justification for believing P.'[1] But if one finds this implication difficult to accept (see McGlynn 2012: s. 4), it is possible to modify the normic theory in such a way as to remove it.

[1] If so, then this constitutes another, quite different, kind of objection to the risk minimisation conception of justification, which straightforwardly predicts that 'justification' should behave much more like 'tall'. On the risk minimisation picture, justification corresponds to a particular threshold on an underlying linear scale.

I have been assuming thus far that, for any non-impossible proposition P, there will be maximally normal worlds at which P holds. With this assumption in place, possible worlds can be assigned *degrees* of abnormality—the most normal worlds will be assigned degree 1, the next most normal worlds will be assigned degree 2, etc. Degrees of abnormality may have a finite maximum, or they may continue indefinitely (perhaps even into the transfinite). This opens up a way of gauging the *strength* of a normic support relation. A body of evidence E normically supports a proposition P just in case E ∧ ~P is a less normal prospect than E ∧ P—just in case the most normal worlds at which E ∧ P hold are more normal than any worlds at which E ∧ ~P hold. But even if this condition is met and E ∧ ~P is a less normal prospect than E ∧ P, there is a further question that might be asked—namely, *how much* less normal? If possible worlds can be assigned numerical abnormality degrees, then this question may permit of a numerical answer.

The larger the normalcy gap between E ∧ P and E ∧ ~P, the stronger the normic support relation between E and P. If the most normal E ∧ ~P-worlds are just 1 degree more abnormal than the most normal E ∧ P-worlds, we can say that E normically supports P to degree 1. If the most normal E ∧ ~P-worlds are 7 degrees more abnormal than the most normal E ∧ P-worlds, we can say that E normically supports P to degree 7. If the most normal E ∧ ~P-worlds and the most normal E ∧ P-worlds are equally normal, we can say that E normically supports P to degree 0. Finally, if there are no possible worlds in which E ∧ ~P holds, then the degree of normic support that E affords P might be regarded as infinite. On the normic theory of justification comparisons, one will have more justification for believing P than Q iff one's evidence normically supports P to a greater degree than Q.

Just as the probabilistic bearing of a body of evidence can be captured by a probability function defined over propositions, the *normic* bearing of a body of evidence, on the present picture, can also be captured by a function assigning numbers to propositions. It's important to note that, without additional assumptions, the numbers used to represent degrees of normic support can be taken to carry only *ordinal* information—nothing entitles us to assume, for instance, that the difference between degrees 1 and 2 equals the difference between degrees 14 and 15 or is smaller than the difference between degrees 20 and 30 etc. The function assigning degrees of normic support to propositions will, as it turns out, meet the formal conditions for a *(completely minimative) positive ranking function*[2] (see Spohn 2012: 75). I will have more to say about the properties of such functions in Chapter 8.

With this framework in place, consider the following proposal: one has justification for believing P iff the degree to which one's evidence normically supports P is greater than a threshold t, where t can be variable and/or vague. This can be thought of as a *threshold* normic theory of justification. It is important to keep in mind that a normic support threshold cannot be understood in quite the same way as an evidential probability threshold—it effectively represents a *position in an ordering*, rather than a cut-off for a quantity. Nevertheless, the threshold normic theory is, in some ways, closer in its form to the risk minimisation theory. On the threshold normic theory, it's clear that one could have more justification for believing P than Q without thereby having justification for believing P—one's evidence could normically support P to a greater degree than Q without either proposition being supported to a degree that exceeds t.[3]

[2] A positive ranking function, like a probability function, is defined over a set of propositions that is closed under negation and disjunction and includes a 'maximal' proposition, entailed by all others in the set. Propositions are standardly modelled, for this purpose, as subsets of a set of possible worlds W, with W playing the role of the maximal proposition. The range of a ranking function is $N \cup \{\infty\}$—the set of non-negative integers plus infinity. A function r taking propositions into $N \cup \{\infty\}$ is a positive ranking function just in case it conforms to the following axioms:

(i) $r(W) = \infty$
(ii) $r(\emptyset) = 0$
(iii) $r(P \wedge Q) = \min(r(P), r(Q))$ for any propositions P, Q

If the domain of the function is infinite and closed under infinite disjunction (conjunction) then (iii) may be strengthened to:

(iv) For any set of propositions Σ, $r(\wedge\Sigma) = \min\{r(P) \mid P \in \Sigma\}$

A ranking function that satisfies (iv) is described as *completely minimative*.

[3] The threshold normic theory of justification will face its own version of the 'threshold problem' mentioned in Chapter 2, n. 2—any normic support value that one chooses is bound to seem arbitrary as a threshold for justification. What I have been calling the normic theory of justification could be thought of as a limiting case of the threshold normic theory, in which the value of t is fixed at 0. Since this is the only threshold that would not seem obviously arbitrary, the normic theory of justification arguably avoids any version of the threshold problem—a feature that sets it apart not only from the threshold normic theory, but from the risk minimisation and hybrid theories as well. Indeed, this may be the only theory of justification that both avoids the threshold problem *and* provides an obvious way of dealing with justification comparisons.

Just as the property of being normically supported is closed under multiple premise deductive consequence, so too is the property of being normically supported to a degree greater than t. As can be easily demonstrated, if my evidence normically supports each of a series of propositions $P_1 \ldots P_n$ to a degree greater than t and $P_1 \ldots P_n$ together entail Q, it follows that my evidence normically supports Q to a degree greater than t. Suppose, once again, that I've invited 100 reliable and trustworthy friends to a dinner party, and all 100 have replied saying that they will attend. If I have justification for believing that Chuck will attend, and for believing that Barbara will attend, etc., then, by the threshold normic theory, each of these propositions must be normically supported to a degree greater than the threshold t. If the value of t is held constant, then the proposition that all 100 invited guests will attend will also be normically supported to a degree greater than t and the threshold normic theory will predict that I have justification for believing it, its low evidential probability notwithstanding. The threshold normic theory, like the simple normic theory, will predict that one can have justification for believing propositions that run a high risk of falsity.

As discussed in section 4.3, one situation in which this prediction seems to come under particular pressure is when the proposition in question assumes some practical importance. Suppose again that I decide to have my dinner party catered and have to instruct the caterers on how much food to prepare. Should I tell them that 100 people will be in attendance, even if they charge substantially more for parties of 100 or larger—even if I could be bankrupted or sued if my estimate turns out to be too high? If I have justification for believing that 100 people will attend then, we might think, it should be rationally permissible for me to tell the caterers this. But, by doing this, I would be running a very high risk of an adverse consequence.

When confronted with cases like this, a threshold normic theorist has a new option, not available to the simple normic theorist—namely, to claim that the value of the threshold t is to be determined in part by one's *practical interests*. On this proposal, the degree of normic support required for one to have justification for believing P may rise when P becomes a matter of practical urgency. Call this the *interest relative* threshold normic theory. For the interest relative threshold normic theorist, the circumstance just described is precisely one that would trigger a rise in the threshold value t—perhaps to the point at which

I cease to have justification for believing that all 100 invited guests will attend. As such, I should presumably *not* tell the caterers to expect 100 people, as I would cease to have justification for expecting this myself.

There need be no failure of multiple premise closure here. When I'm speaking to the caterers, the threshold t may also rise for the propositions that Chuck will attend, that Barbara will attend, etc. in which case I will also lose justification for believing these propositions and be obliged, perhaps, to fall back upon propositions that are probabilistically qualified—Chuck will *probably* attend, Barbara will *most likely* attend, etc.[4]

In section 4.3 I made the following claim about the relationship between justification and actionability: if I have justification for believing a proposition P then it should, in some circumstances, be rationally permissible for me to act on the assumption that P is true—to use P as a premise in my practical reasoning, and choose actions that will lead to beneficial consequences if P is true. The claim is highly qualified—and the reason for this is that whether it is rationally permissible to act upon a proposition P is not just a function of the strength of one's evidence for P, but also depends on one's practical interests. Speaking generally, the more that is at stake over a given decision—the more one stands to lose if one acts on a proposition that turns out to be false—the greater the degree of justification one requires for that proposition in order to be rationally permitted to act upon it.

If a great deal is at stake over the decision of what to tell the caterers, then it may not be rationally permissible for me to act on the assumption that all 100 guests will attend. As I suggested, however, this need not imply that I lack justification for believing this proposition. What it implies is that, given how much is at stake, my justification for believing this proposition is not of a *sufficiently high degree*.

If we accept the interest relative threshold normic theory described here, then we can embrace an unqualified principle linking justification to actionability: if I have justification for believing a proposition P then it will be rationally permissible, in any circumstance, for me to act on the

[4] Suppose that, while I'm speaking to the caterers, one of the invited guests calls up and cancels. In this case my justification for believing, of each other guest, that he or she will attend would, arguably, be immediately restored. At this point, there would no longer be any question of telling the caterers to prepare for 100 and, as such, I would no longer be facing a high stakes decision. This would be an odd result of the present theory—perhaps one of many odd results in the vicinity. Thanks here to Kenny Easwaran.

assumption that P is true. The price for this is that whether one has justification for believing a proposition P will no longer be determined solely by the strength of one's evidence for P—one's practical interests will also be relevant. On this theory, there could be two subjects with the very same evidence, providing equally strong support for P, who, owing to their different interests, have a different justificatory status with respect to P. Some epistemologists have embraced theories of justification that share this consequence (Owens 2000, Fantl and McGrath 2002, 2009).[5]

Whether one is rationally permitted to act upon a proposition P depends both on how strongly one's evidence supports P and on one's practical interests. Given this common ground, there are effectively two ways in which we can fit the notion of justification into the picture. First, we can maintain that whether one has justification for believing P is solely determined by the strength of one's evidence for P, but whether one is rationally permitted to act upon P is not solely determined by whether one has justification for believing P. Second, we can maintain that whether one is permitted to act upon P is solely determined by whether one has justification for believing P, but whether one has justification for believing P is not solely determined by the strength of one's evidence for P. My intention here is not to endorse either of these approaches, but to emphasise that both can be accommodated within a normic framework for understanding justification. The simple normic theory and the non-interest relative threshold normic theory will take one down the first, more traditional, path, while the interest relative threshold normic theory will take one down the second path.

5.3. Four Theories

Using the notion of normic support, I have described three possible theories of justification. According to the hybrid theory, one has justification for believing P iff (i) one's evidence normically supports P and (ii) the probability of P, given one's evidence, exceeds a threshold t (where t is a real number close to, but less than, 1). According to the normic

[5] Brian Weatherson (2005, 2011) has defended an interest relative account of justified belief where the notion of belief, rather than the notion of justification, is the source of interest relativity. These theories also bear a relation to interest relative invariantist theories of knowledge attributions, such as those defended by Hawthorne (2003) and Stanley (2005).

theory, one has justification for believing P iff one's evidence normically supports P. According to the threshold normic theory, one has justification for believing P iff the degree to which one's evidence normically supports P exceeds a threshold t (where t is a positive integer). A fourth theory, too, is possible—another sort of hybrid that combines risk minimisation with a threshold-based normic support requirement: one has justification for believing P iff (i) the degree to which one's evidence normically supports P exceeds a threshold t_1 and (ii) the probability of P, given one's evidence, exceeds a threshold t_2 (where t_1 is a positive integer and t_2 is a real number close to, but less than, 1).

I noted in the introduction that a handful of vague slogans about justification does not amount to a competitor to a relatively clear and detailed theory such as the risk minimisation conception. But the theories that I've presented here—successors to those slogans—should, I think, be considered genuine competitors. It is true, of course, that I haven't supplied a full philosophical analysis of the notion of normic support—in some sense this notion has been left mysterious. But, by the same token, proponents of the risk minimisation conception have not supplied full philosophical analyses of the notion of evidential probability. This notion too is mysterious in a sense—and we shouldn't let its familiarity as a philosophical tool lull us into thinking otherwise.

Epistemologists have, of course, offered *accounts* of evidential probability—such as the internalist and externalist accounts briefly discussed in section 2.1. But these are just pictures, painted with a broad brush, of the *sort of thing* that evidential probability is supposed to be—they don't come anywhere close to philosophical analyses, in the sense of precise, informative necessary and sufficient conditions for a proposition to have a certain probability value, given a particular body of evidence.[6] I, too, have provided a broad brush picture of the sort of thing that I take normic support to be—one that, I'm willing to admit, doesn't come anywhere close to a full philosophical analysis as traditionally conceived.

[6] It is perhaps Carnap (1950) who comes closest to offering something like a philosophical analysis of the notion of evidential probability. Carnap's analysis is now widely regarded as inadequate, partly for its reliance on the idea that probabilistic relations between sentences could be a pure function of their syntax. Subsequent philosophers have, by and large, been much less ambitious when speculating about the nature of evidential probability.

It may well be that neither of these notions permits of a full philosophical analysis. Some have suggested that we should be content to treat evidential probability as something like a primitive in epistemology—that our pretheoretic grasp of the notion is good enough for it to be put to epistemic work, without our insisting on a precise definition (Keynes 1921: chap. 1, s. 8, Williamson 2000: s. 10.1). I am not unsympathetic to this idea—but I think we would be deluding ourselves if we thought that our pretheoretic grasp of the notion of evidential probability was somehow firmer than our pretheoretic grasp of a notion like normalcy. If evidential probability makes for a respectable primitive then so too does normic support.

While we lack a philosophical analysis of the notion of evidential probability we do, of course, have a systematic theory of its formal behaviour—and this, I imagine, plays no small part in explaining our relative comfort with it. But the notion of normic support is also perfectly amenable to systematic formal treatment—and the formal treatment of normic support, like the formal treatment of evidential probability, turns out to be quite unhampered by the fact that there may be unanswered philosophical questions about the notion. In this chapter, I have already outlined some of the rudiments of a formal treatment of normic support—and it will be further explored and developed over the forthcoming chapters.

6

Protection from Error

6.1. Safety

In this chapter I shall approach the normic support condition on justification from a different angle, portraying it as the close analogue of a widely accepted condition on knowledge—the condition of *safety*. A number of epistemologists have been attracted to the idea that, in order to qualify as knowledge, a belief must be appropriately protected from the possibility of error—it must be the case that the belief could not easily have been false, given the way that it was formed and the evidence upon which it was based (Williamson 1992, 2000: chap. 5, Sosa 1999a, 1999b, Pritchard 2005: chap. 6, Smith 2009). Safety conditions are, in effect, attempts to develop this idea and make it more precise. In section 1.3, I introduced safety in this way: suppose one believes P based upon evidence E. E might be described as safe evidence for P iff P could not easily have been false, given that E is true. Or, in possible worlds parlance, E might be described as safe evidence for P iff in all very close or similar possible worlds in which E is true, P is true. For E to be safe evidence for P, E and P must dovetail throughout a certain set of significant possible worlds. One situation of this kind is depicted in Figure 6.1, where w is the world at which one believes P based on E, the inner circle represents the set of worlds that are very similar to world w and the outer ring represents the set of worlds that rank next in terms of similarity to w.

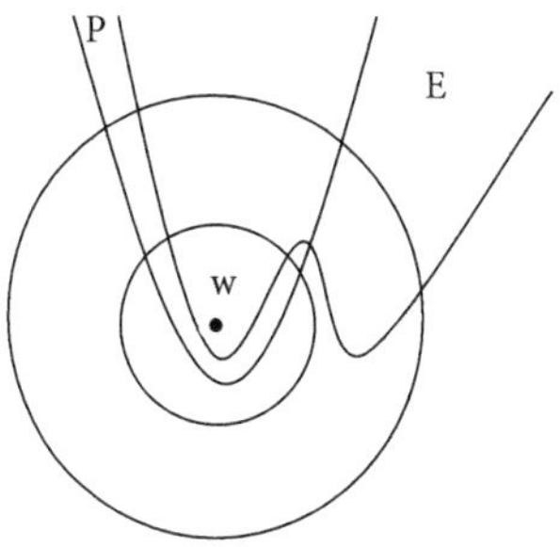

Figure 6.1.

A belief that is based upon safe evidence might be described as a *safe belief*.

I will assume that, if one believes a proposition P based upon evidence E, then E must be a part of one's body of evidence. I will continue to assume that one's body of evidence can consist only of truths. From these assumptions it follows that safe beliefs must be true beliefs. If E is safe evidence for P at w, then P is true at all E-worlds that are very similar to w. Since w is an E-world and, presumably, counts as very similar to itself, P must be true at w. Put in a slightly different way, if a subject believes P based on E at w, then the E-region in the space of possible worlds must include w. If E is safe evidence for P then, within the inner circle, the E-region must be contained in the P-region. Since w itself resides within the inner circle, it follows that the P-region must include w.

The idea that possible worlds stand in relations of comparative similarity and difference was introduced into the philosophical mainstream by David Lewis in *Counterfactuals* (1973). Lewis originally invoked the notion of world similarity as a part of his semantic analysis of counterfactual conditionals. His idea, roughly put, was that a conditional such as 'If it were not raining today, then we would have gone to the cricket' means something like: in any possible world in which the weather is fine and which is as similar as the weather being fine permits it to be, we go to the cricket.

Let w be the index world at which a counterfactual conditional is to be evaluated. If we visualise possible worlds arranged in space, with proximity serving as a metaphor for similarity, then we can imagine worlds falling into a series of concentric spheres radiating out from w, as in Figure 6.1. The innermost sphere will contain those worlds that are most similar to w, the next sphere will incorporate worlds that are almost as similar but not quite, and so on. In the mathematician's sense, a sphere is defined as a set of points (in three-dimensional space) equidistant from a single point. Lewis, though, wants us to understand spheres as being *solid regions* rather than hollow shells—each sphere includes the worlds in previous spheres but adds more.

According to Lewis, the largest similarity sphere around w represents the set of worlds that are possible from the perspective of w. At the opposite extreme the smallest non-empty similarity sphere[1] around a

[1] Lewis stipulates that the empty set always counts as one of the spheres around a world w. The inclusion of an empty sphere proves technically convenient, but makes no difference when it comes to evaluating the truth of counterfactuals.

world w will, according to Lewis, be a set containing just w itself—a constraint he termed (strong) *centering*. Presumably no world resembles w more closely than, or as closely as, it resembles itself. Say that a sphere around a world w *permits* a proposition just in case it contains worlds at which the proposition is true. In order to determine whether a counterfactual conditional holds true at w, we travel out from w, sphere by sphere, until we encounter worlds in which the antecedent holds. If all of the antecedent-worlds subsumed in this sweep are also consequent-worlds, then the conditional will be true at w—otherwise it will be false.

A little more formally, if we introduce a counterfactual conditional operator □→ to be read 'If __ were the case, then . . . would be the case', it might be given the following truth condition:

> P □→ Q is true at w iff there exists a P-permitting similarity sphere around w in which all of the P-worlds are Q-worlds.

Lewis allows for one exception to this: P □→ Q should be deemed vacuously true at w if there is *no* P-permitting sphere of similarity around w.

On Lewis's account, counterfactual conditionals work in much the same way as strict conditionals—namely, as a kind of universal generalisation about a class of possible worlds. But while the quantificational range of a strict conditional, at a given world, is constant, the quantificational range of a counterfactual conditional, at a given world, *varies* as a function of the antecedent. The similarity spheres around that world provide the various possible quantificational ranges, with a counterfactual conditional quantifying over the smallest similarity sphere to accommodate the antecedent's truth. It is for this reason that Lewis described counterfactual conditionals as *variably strict*.

Most epistemologists who have proposed safety conditions upon knowledge have taken themselves to be working with the same basic notion of world similarity as Lewis. Indeed, some early proponents of safety conditions tended to formulate them, first, as counterfactual conditionals, with the possible worlds formulation coming only via a Lewis-style truth condition (Sosa 1999a, 1999b). There are some historical reasons for this—some early proponents of safety saw the condition as a modification of Nozick's sensitivity condition—which *is* very naturally formulated as a counterfactual conditional. Nevertheless, I'm inclined to think that this formulation of safety gets things backwards.

The possible worlds formulation, for those accustomed to thinking in such terms, captures a safety-type notion in a very transparent way—and its viability should not be made to depend upon the viability of a certain truth condition for counterfactual conditionals.

If we do adopt something approaching the Lewisian picture, however, then the safety condition that I've outlined here could be reformulated as a counterfactual conditional: E is safe evidence for P just in case E □→ P. Importantly, though, this reformulation only works if we relax Lewis's strong centering constraint and replace it with another constraint that he calls *weak centering*. In a strongly centered system of spheres, as mentioned, the index world will be the only world in the smallest non-empty similarity sphere around it. In a weakly centered system of spheres, while the index world is still guaranteed to be a member of the smallest non-empty similarity sphere, other worlds are permitted to infiltrate.

Suppose one believes P based on E at world w. If strong centering is in place then {w} will be the smallest non-empty similarity sphere around w. Since E must be true at w, {w} will be E-permitting. It follows that E □→ P will be true at w on the condition that P is true at w. The conditional formulation of safety would predict that E is safe evidence for P whenever P is *true*. If we replace centering with weak centering, however, then the smallest E-permitting similarity sphere around w may contain other worlds, aside from w. In this case, E □→ P will be true at w iff all of the E-worlds in this sphere are P-worlds. If we interpret this smallest similarity sphere as comprising those worlds that are 'very similar' to w, we will have arrived at the familiar possible worlds definition of safe evidence. With weak centering in place, then, we can say that E is safe evidence for P at w just in case E □→ P is true at w or, in possible worlds parlance, just in case there is an E-permitting sphere of similarity around w in which all of the E-worlds are P-worlds.

One way to motivate the idea that safety is necessary for knowledge is by considering examples, such as Ginet's well-known Gettier case (first described in Goldman 1976): suppose that I'm driving through the countryside, stop my car in front of a barn and come to believe that there is a barn before me based upon the evidence that there appears to be a barn before me. In ordinary circumstances, such a belief would surely qualify as knowledge and would also be safe—in all very similar worlds in which there appears to be a barn before me, there is a barn before me. But now suppose that, unbeknownst to me, the surrounding area is peppered with cleverly constructed paper mache barn façades that cannot be distinguished from real barns when viewed from the road. It seems that my belief should not qualify as knowledge in this case, and would not be safe either—there would be very similar possible worlds in which there appears to be a barn before me even though the object before me is a façade.

Additional support for a safety condition on knowledge comes from observing certain conversational patterns. A possible way of objecting to a knowledge claim is to say something along the lines of 'You don't know that—you could easily be wrong'. The '...you could easily be wrong' is naturally understood as suggesting that the belief in question violates a safety condition of the sort described here. Indeed, one quite general way to challenge the propriety of a belief or assertion is to allege that it could easily be false. If safety were not necessary for a belief to qualify as knowledge, it's difficult to see why this should be effective as a challenge. Not all epistemologists of course are moved by these considerations—and some have denied that there is a safety condition on knowledge (see Neta and Rohrbaugh 2004, Comesaña 2005). I won't attempt to give an extended defence of the condition here—at least not of a standard sort. Another, rather different, kind of argument for a safety condition on knowledge involves showing just how much *can be done with it*—how much our epistemology can be enriched by its incorporation. This chapter, along with Chapters 7 and 8, might be understood as providing an argument of this kind.

6.2. Rediscovering Normic Support

As I noted at the beginning of the chapter, a number of epistemologists have been attracted to the idea that, in order to qualify as knowledge, a belief must be appropriately protected from the possibility of error. But there is something to the thought that, in order to qualify even as *justified*, a belief must be appropriately protected from the possibility of error. This idea cannot, however, be developed in quite the same way as it was in the case of knowledge. In order to qualify as justified, a belief is not required to meet a safety condition of the sort described in section 6.1. One immediate reason for this is that safety is a *factive* property whereas justification is not—safe beliefs must be true, while justified beliefs can be false. Whatever else it might require, justification is consistent with *actual* error. Perhaps this has led some to think that justification could not possibly demand any kind of *protection* against error—minimisation of the chances of error is the most that it could require. If some have thought this, then they have been too quick—and have been led too quickly to embrace a picture of justification that makes it into a very different sort of thing from knowledge.

The fact that safety demands truth is in fact symptomatic of a more general feature of safety that seems not to be shared by justification—whether one's belief counts as safe is something that is highly dependent upon the nature of the environment in which one finds oneself. In Ginet's case, the presence of barn façades in my vicinity is enough to prevent my belief that there's a barn before me from qualifying as safe or as knowledge. But it would not prevent my belief from qualifying as justified. Provided I'm oblivious to the presence of the façades, I would be perfectly justified in believing that there is a barn before me—it is partly for this reason, of course, that the barn-façade example has been thought to count as a Gettier case. Safety, in a manner of speaking, is a *modally fragile* condition—it doesn't take much of a change in the external circumstances to compromise the safety of a belief. Justification, on the other hand, appears to be more *modally robust*—if we hold fixed a subject's belief, and a subject's evidence and the way the belief is based upon that evidence, then the justificatory status of that belief seems very resilient to changes in the external environment, and perhaps even completely resilient to any such changes (more on this below).[2]

Even if the *very object* that I stopped in front of was a barn façade, and my belief was consequently false, this would be yet another environmental change that would not disturb its justificatory status. If we hold fixed a subject's belief and the evidence on which it is based, it may take only a small change in the external environment to transform that belief from a true one into a false one. The fact that false beliefs can count as justified is, in a way, just a side effect of the modal robustness of justification. Any error protection condition upon justification would, then, have to be a condition that is consistent with actual error, and a condition that is

[2] All Gettier cases, in effect, serve to illustrate the comparative modal robustness of justification as opposed to knowledge. Indeed they illustrate more than this—that even the conjunction of justification and truth exhibits a greater modal robustness than knowledge does.

modally robust. Formally, at least, it is straightforward to construct an error protection condition with the required features.

Another formal constraint upon sphere systems that Lewis considers is a constraint he terms *absoluteness*, according to which all possible worlds are assigned the *same* system of spheres. With an absoluteness constraint in place, there is just one 'true' sphere system with which every world is associated. If we impose this constraint, an error protection condition would enjoy maximal modal robustness—whether or not it is met at a given world would be completely independent of the nature of that world. Suppose one believes P on the basis of evidence E at world w. If there is an E-permitting sphere associated with w in which every E-world is a P-world, then, under absoluteness, this sphere must be associated with every world. Similarly, if there is no E-permitting sphere associated with w in which every E-world is a P-world, then, under absoluteness, no such sphere is associated with any world. Under absoluteness, the spheres don't change as we move from world to world.

In a weakly centered system of spheres, recall, the index world must be a member of the smallest non-empty sphere associated with itself. Absoluteness is incompatible with weak centering—or, more precisely, the two constraints are compatible only in the degenerate case in which every possible world is assigned the totality of possible worlds as its sole sphere (Lewis terms this *weak triviality*). In what Lewis termed a *decentered* system of spheres, the index world enjoys no particular pride of place—it need not be a member of the smallest non-empty sphere[3] and can appear anywhere in the system. If we set aside the degenerate case, then embracing absoluteness obliges us to give up weak centering and allow for decentered sphere systems.

Further, once we abandon weak centering, our error protection condition would no longer require truth. Suppose again that one believes P on the basis of evidence E at world w. Even though E must be true at w, if the system of spheres associated with w is decentered, then there is no guarantee that w is one of the E-worlds in the smallest E-permitting sphere. As such, even if there is an E-permitting sphere associated with w

[3] Once weak centering is relaxed there are, strictly, no constraints in place that guarantee the existence of a smallest non-empty sphere in a system—we might have an infinite chain of ever smaller spheres that continues without end. I will ignore this complication for the time being, but will return to it in Chapter 8.

in which every E-world is a P-world, P need not be true at w. Figure 6.2 depicts one kind of situation in which this can arise:

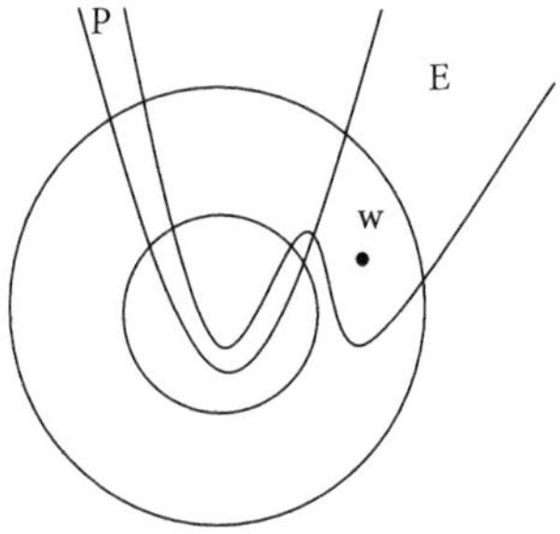

Figure 6.2.

If a subject believes P based on E at w then, as before, the E-region must contain w. Even if it's true that, within the inner circle, the E-region is contained in the P-region, since w need no longer reside in the inner circle, it may still lie outside of the P-region.

Decentered sphere systems cannot, of course, be taken to represent information about comparative similarity to the index world, and the error protection condition that is now being modelled could not be a safety condition as such. I suggested in section 2.4 that possible worlds may be compared not only for their similarity, but also for their *normalcy*. Information about the comparative normalcy of possible worlds, from the perspective of an index world w, could easily be represented by using a system of spheres—with the innermost sphere containing those worlds that are most normal from the perspective of w, the next sphere incorporating those worlds that are almost as normal but not quite, and so on. The system of spheres will be weakly centered on the assumption that the index world estimates itself to be maximally normal. Since this assumption will not, in general, be met, such sphere systems will, in general, be decentered.

The question of what it takes for one world to count as more normal than another is one that has already received some discussion. But, to come at it with somewhat fresh eyes, one potentially helpful way to think about normal worlds is by analogy with *idealised models*. The practice of using simplified, idealised models of potentially complex actual phenomena is one that is very familiar to us—the idealised models of projectile motion routinely used in mechanics provide one vivid illustration. We cannot explain everything all at once—we have to abstract away from, or control for, certain factors in order to expose underlying patterns and mechanisms (Pietroski and Rey 1995: s. 2.1, see also Bhaskar 1975: chaps 1 and 2). One way to approach the idea of a 'normal world' is as a kind of idealised model *writ large*. On this sort of picture, worlds will count as normal, from w's perspective, to the extent that they approximate simplified, idealised models of w. And w need not in general serve as a good idealised model of *itself*. This is reminiscent of the lesson sometimes taken from Borges's 'On exactitude in science'—a map loses its usefulness once it becomes too large, too detailed, and too accurate (Borges 1935).[4]

In the most normal worlds in which there appears to be a barn before me, there is a barn before me. And this will *not* depend in any way on whether there are barn façades in the actual vicinity—or even if the very object before me happens to be a façade. Even if there is a façade before me, and other façades in the vicinity, the most normal worlds in which there appears to be a barn before me won't be like this—they will be idealised in appropriate respects.

If decentered sphere systems are interpreted as carrying information about the comparative normalcy of worlds, then the error protection condition defined in terms of such spheres will simply be the condition of normic support—in just the sense introduced in section 2.4. As defined there, E normically supports P at w just in case there is a possible world in which E and P are both true and which is more normal, from the perspective of w, than any world in which E is true and P is false. This is just another way of saying that there is an E-permitting sphere of

[4] A very different way of understanding world normalcy is suggested by Brian Weatherson (MS, s. 4). Weatherson suggests that a world w_1 should count as more normal than a world w_2, given a body of evidence E just in case w_1 is more similar to E, in something like Lewis's sense, than w_2. If E were a complete catalogue of truths about the actual world then world normalcy and world similarity might be expected to coincide, but for incomplete bodies of evidence, they may diverge radically. One thing that is striking about Weatherson's proposal is that it makes world normalcy into an *evidence relative* notion—we cannot ask anymore whether one world is more normal than another per se, but only whether one world is more normal than another, given a particular body of evidence. On this proposal, a subject's body of evidence will do a kind of double duty in determining which propositions are normically supported for that subject—first, by sorting worlds into normalcy spheres and, second, by eliminating those worlds in which it doesn't hold. The normically supported propositions will be those that are true at all the uneliminated worlds in the smallest uneliminated normalcy sphere. This is not the approach pursued here, though I think it would be a very interesting idea to explore—and one that will change the formal behaviour of normic support in a range of ways.

normalcy associated with w in which all of the E-worlds are P-worlds. What it is for E to normically support P is for E and P to dovetail throughout a certain set of significant possible worlds. But, rather than selecting these worlds on the basis of how similar they are, as we would if we were evaluating the safety of E as evidence for P, we select them instead on the basis of how normal they are. What it is for a *belief* to be normically supported, then, is for it to be based upon evidence that normically supports its content.

Normic support, like safety, represents a kind of protection against the possibility of error. If a belief is safe, then the belief could not easily have been false, given the way that it was formed and the evidence upon which it was based. If a belief is normically supported, then the belief could not *normally* have been false, given the way that it was formed and the evidence upon which it was based.

If we model safety in terms of possible worlds, as is done here, then it will turn out that any evidence will be safe evidence for a necessary truth. More generally, it will turn out that necessarily equivalent propositions must always share the same safety profile, given the same evidence. Some epistemologists have taken the step of exempting necessary propositions from their preferred analyses of safety (Pritchard 2005: chaps 3 and 6)—a measure which may help with the former problem, though not with the latter more general one. If we model normic support in terms of possible worlds then, as briefly noted in section 2.4, analogous problems will arise—any evidence will normically support a necessary truth and necessarily equivalent propositions will always have the same normic support profile, given the same evidence.

These results might seem particularly troubling if we consider a posteriori necessary truths or propositions that are necessarily equivalent a posteriori. It seems clearly wrong, for instance, that the proposition expressed by 'Cicero is Tully' should be normically supported by any evidence whatever—this, after all, seems to be precisely the sort of thing on which evidence ought to bear in a non-trivial way.[5] We can avoid this

[5] If propositions are taken to be appropriately *coarse-grained* then this need not, however, be so. Direct reference theorists, insofar as they continue to work within the framework of propositions, and accept a plausible compositionality thesis, must take 'Cicero is Tully' to expresses the *very same proposition* as 'Cicero is Cicero' (for some discussion see Soames 2002: chaps 1 and 8, 2011). But then it seems correct to say that any body of evidence will normically support *this proposition*. On this picture, having justification to believe a proposition will be

consequence if the worlds that feature in the formal characterisation of normic support are understood not as *metaphysically* possible worlds—worlds that really could have obtained—but, rather, as *epistemically* possible worlds—worlds that could coherently be conceived to obtain. While it may be metaphysically necessary that Cicero is Tully, this is not epistemically necessary (see, for instance, Kripke 1980, Soames 2011). If we characterise normic support in terms of epistemically possible worlds, the proposition expressed by 'Cicero is Tully' need not be normically supported, even though this is a metaphysically necessary truth, and need not share the normic support status of the proposition expressed by 'Cicero is Cicero', even though the two are equivalent as a matter of metaphysical necessity.

Epistemically possible worlds are often understood as being *centered* on a designated individual at a given time (see for instance Chalmers 2011: s. 4). To conceive of a world as obtaining, it's not enough to imagine certain qualitative facts to hold—one must also imagine *oneself* to have a particular identity and location within that world. The individual and time upon which a world is centered serve to specify just who one *is* within that world. The use of centered worlds will also serve to make relations of normic support potentially more discriminating. If the worlds that feature in the formal characterisation of normic support are understood as being centered, then the propositions that Bruce expresses by 'I live in Australia' and 'Bruce lives in Australia' will have different possible world truth conditions and could have different normic support profiles (for discussion of centered worlds and centered propositions see Lewis 1979: s. X, Egan 2006: s. 3, and Huber 2009: s. 2).

Logical truths and logical equivalences are, of course, generally regarded as being epistemically, as well as metaphysically, necessary. As such, we are still left with the result that logical truths will be normically supported by any body of evidence and that logically equivalent propositions will share the same normic support status, given the same evidence. We might avoid even this result if we were prepared to

neither necessary nor sufficient for having justification to accept a particular *sentence* expressing it. We might wonder just what is required for the latter—but it's not clear that it falls to the present theory to supply an answer to such a question. These considerations may not apply so straightforwardly to other putatively a posteriori necessary truths and falsehoods, such as 'Gold is an element' and 'Gold is a compound'. I assume in the body text that propositions are relatively fine-grained.

allow *impossible worlds* to enter into the formal characterisation of normic support (see, for instance, Rantala 1982). On reflection, though, I am unsure whether this result ought to be avoided.

Even if we regard normic support as both necessary and sufficient for justification, all that will follow is that one must always have justification for believing any logical truth and that logically equivalent propositions must always have the same justificatory status. It will not follow that a *belief* in a logical truth must always be a justified belief or that *beliefs* in logically equivalent propositions must always have the same justificatory status. In order for a belief to be justified it is necessary that it be properly based upon one's evidence—and the conditions on proper basing may be far from trivial.[6] Neither will it follow that we are under any epistemic *obligation* to believe logical truths or to believe logical equivalents of propositions we already justifiably believe. On the contrary, we may be epistemically obliged to give such beliefs up, if they are improperly based.

In any case, the claim that one must always have justification for believing a logical truth and that logically equivalent propositions must always have the same justificatory status may be a consequence of treating normic support as necessary and sufficient for justification—but it is also a simple consequence of the principle of (single premise) closure. This will be a feature of *any* theory of propositional justification that respects the principle.[7, 8]

[6] In particular, in order to properly base a belief in P upon one's evidence E it may be necessary that one be appropriately sensitive to the fact that E provides justification for believing P. If P is a complicated logical truth and E is a completely unrelated truth then, in order to be appropriately sensitive to the fact that E provides justification for believing P, one may already need some independent appreciation for the fact that P is logically true. Similarly, if one's evidence E provides clear justification for a proposition P but not for a complicated logically equivalent proposition Q then, in order for one to be appropriately sensitive to the fact that E provides justification for believing Q, it may be necessary that one have some independent appreciation for the fact that P and Q are logically equivalent. I am indebted here to conversations with Conor McHugh—and see McHugh: 2014: s. VIII.

[7] Provided, at any rate, that the background logic supports the 'implosion' scheme, according to which a logical truth will deductively follow from any premise: if Q is a logical truth then $P \vdash Q$. This scheme is valid in classical and intuitionistic logic amongst others. If we were working in a relevant logic in which implosion failed, then we would anyway be under significant pressure to modify the formal treatment of normic support—incorporating logically impossible worlds or using some equivalent device to alter its formal properties accordingly.

[8] In Chapter 2 n. 6 I set out a closure principle for *doxastic* justification along the following lines: if one justifiably believes each of a set of propositions, and deduces a further proposition from them, then this further proposition is justifiably believed. A principle like

Suppose we insert the notion of world normalcy into a Lewis type truth condition in order to define a new conditional operator $\blacksquare\!\!\rightarrow$:

> P $\blacksquare\!\!\rightarrow$ Q is true at w iff there exists a P-permitting normalcy sphere associated with w in which all of the P-worlds are Q-worlds.

We can allow for the same sort of exception that we acknowledged in the case of the counterfactual conditional—P $\blacksquare\!\!\rightarrow$ Q should be deemed vacuously true at w if there is no P-permitting sphere of normalcy associated with w.

One plausible natural language interpretation of this operator is as a kind of *normic conditional* (see Scriven 1959, Bhaskar 1975, Schurz 2001)—a conditional in which the connection between antecedent and consequent is mediated by a qualifying normalcy or ceteris paribus clause. In line with this suggestion, $\blacksquare\!\!\rightarrow$ could be read 'If __ were the case then normally . . . would be the case' or 'If __ were the case then ceteris paribus . . . would be the case' (Smith 2007).[9] It is widely thought that adding a ceteris paribus clause to a counterfactual conditional serves to strengthen the antecedent—albeit in a rather vague and unspecified way (see, for instance, Schiffer 1991, Hempel 1988, Earman and Roberts 1999, Earman, Roberts, and Smith 2002). On the present approach, though, the ceteris paribus clause should be assimilated not to the antecedent of the conditional but, rather, to the conditional operator itself (compare Bhaskar 1975: chap. 2, s. 4). Adding a ceteris paribus clause to a counterfactual may well alter the set of possible worlds that proves relevant for its evaluation. But, on the present approach, it does

this will not make it trivial to justifiably believe a logical truth. If R is a logical truth and P, Q . . . are unrelated propositions that I happen to justifiably believe then 'deducing' R from P, Q . . . will effectively involve proving it from no premises whatsoever. Similarly, if R is a very complicated logical equivalent of $P \wedge Q \wedge \ldots$ then deducing it from P, Q . . . will involve proving that this is so.

[9] Boutilier (1994) proposes a comparative normalcy possible world semantics for what he terms 'normative' conditionals—that is, conditionals of the form 'If__, then normally . . .'. He suggests that a normative conditional should be deemed true at a possible world w iff for every possible world at which the antecedent is true and the consequent false, there is a more normal possible world at which (i) antecedent and consequent are both true and (ii) at all possible worlds that are more normal still, if the antecedent is true, then the consequent is true (Boutilier 1994: 103). Boutilier's semantic apparatus is more austere than mine, consisting of a class of possible worlds and a single ordering relation upon that class intended to represent a relation of comparative normalcy. As such, his analysis effectively builds in the presumption that standards of normalcy are world absolute.

this not by modifying the antecedent of the conditional but, rather, by modifying the criteria by which the relevant antecedent-worlds are selected—signalling that they are to be selected on the basis of their normalcy, from the perspective of the index world, rather than their similarity to it.

Adding a normalcy or ceteris paribus clause to a conditional is a widely recognised technique for insulating the conditional against potential refutation. By adding such a clause I make the truth of the conditional less dependent upon prevailing circumstances—and perhaps independent of them altogether. In the terminology that I've introduced, by adding a ceteris paribus clause to a conditional I make its truth more *modally robust*. Suppose that, unbeknownst to me, the power to my house has just been cut off. The straight counterfactual 'If I were to switch on the kettle, it would soon boil' is presumably false. But the normic conditional 'If I were to switch on the kettle then ceteris paribus it would soon boil' is arguably *true*. Although the power has been cut off in the actual world, the most normal worlds in which I switch on the kettle won't be like this. The truth of the straight counterfactual conditional is hostage to whether or not there is a power outage. By adding the ceteris paribus clause one can make the truth of the conditional independent of this factor, amongst others.[10]

Just as safety could be defined using the straight counterfactual conditional, normic support could be defined using the normic conditional: suppose one believes P on the basis of E. E will be safe evidence for P just in case E □→ P and it will normically support P just in case E ■→ P. If this is right, then the safe evidence relation can be thought of as a special case of the counterfactual conditional relation and the normic support relation can be thought of as a special case of the normic conditional relation.

[10] Priest (2001: chap. 5) suggests that straight counterfactual conditionals already contain tacit normalcy or ceteris paribus qualification (see also Gundersen 2004). But, as the example in the text illustrates, counterfactuals and normic conditionals are clearly distinct. Saying that if P were the case then Q would be the case is one thing—saying that if P were the case then ceteris paribus Q would be the case is another. One further example: suppose Sam suffers from motor neurone disease. In this case the straight counterfactual 'If I were to strike Sam's patellar tendon then his leg would jerk' may well be false. However, the normic conditional 'If I were to strike Sam's patellar tendon then ceteris paribus his leg would jerk' seems true. After failing to elicit the knee-jerk response I could, for instance, utter this conditional precisely as a way of drawing attention to Sam's condition.

When one first encounters the notion of comparative world normalcy, it's tempting, as noted above, to demand some principled way of determining when one possible world should count as more normal than another. But we would be reluctant, I think, to allow a corresponding demand to waylay the use of a notion like world similarity. And yet, the notion of world similarity is, in its own way, every bit as opaque as the notion of world normalcy is. There are any number of different respects in which possible worlds might be thought to resemble or be similar to one another—and it's only by assigning some system of weights to these different respects that one could possibly arrive at an *overall* measure of world similarity of the sort that Lewis invokes. What is opaque, at least at first blush, is just how this weighting ought to be done.

For Lewis, we don't need to settle this matter as a prerequisite to appealing to the notion of world similarity in furnishing counterfactuals with truth conditions. On the contrary, it is intuitions about counterfactuals *themselves* that will help to show us, via the proposed truth condition, just what sorts of features are important for determining the comparative similarity of possible worlds. Lewis's theory may give us a way of translating any counterfactual conditional into a non-conditional claim about the comparative similarity of possible worlds—but the ambition is not to try and form some independent judgement about the world similarity claim in the hope of arriving at a verdict about the counterfactual. In general, the very opposite will be true—our verdict about the counterfactual will guide our judgement about the comparative world similarity claim.

My attitude toward the notion of comparative world normalcy is a similar one. I think it would be a mistake to try and build some theory of world normalcy from first principles. Rather, normic conditionals may provide us with one way of better understanding the notion, via the kind of truth condition outlined above. The truth condition won't, in general, help us to determine whether particular normic conditionals are true or false. That's not to say that the truth condition is untestable or that it makes no predictions whatsoever. Treating normic conditionals as involving variably strict quantification over possible worlds makes definite predictions about the *logic* to which such conditionals will be subject. Perhaps the most tangible benefit of a semantic analysis such as this is that it allows us to approach questions of logic in a more systematic, less piecemeal way. I will return to this in Chapter 7.

6.3. An Application—Epistemic Contextualism

The claim that justification requires normic support is, in a sense, the justification-theoretic equivalent of the claim that knowledge requires safety. It may be true, as a more general rule of thumb, that the notions of normic support and world normalcy can do much the same work, when theorising about justification, that the notions of safety and world similarity do when theorising about knowledge. In this section I will give one illustration of how this might work. Keith DeRose (1995, 2004) has defended an influential version of contextualism about knowledge attributions that makes extensive use of the notions of safety and world similarity. Using the notions of normic support and world normalcy it may be possible to develop a corresponding contextualist theory of justification attributions.

Contextualism about knowledge attributions, simply put, is the view that sentences of the form '. . . knows that__' are context sensitive in that they express different truth conditions in different contexts of use. Furthermore, according to contextualists, this context sensitivity owes specifically to the presence of the verb 'knows' and will remain even when all extraneous context sensitive vocabulary has been controlled for. One way of developing this idea is to posit a range of *epistemic standards* that can feature in the truth conditions of a knowledge attribution. Such standards are usually taken to be at least partially ordered depending upon how demanding they are—at one end of the spectrum are stringent standards that it is very difficult for a subject to meet, and at the other end are lenient standards that it is relatively easy for a subject to meet. A knowledge-of-P attribution will be true only if the subject meets epistemic standard s with respect to P, where s is whatever standard is appropriate given the attributor's context. The more stringent this standard, the more difficult it will be for the subject to meet it and the more difficult it will be for the attribution to express a truth. As the attributor's context varies, so too do the truth conditions of the attribution.[11]

[11] Some contextualists regard the verb 'knows' as a kind of indexical expression—literally picking out different relations relative to different contexts of use in much the same way that expressions such as 'I', 'here', and 'now' pick out different individuals, places, and times (Cohen 1988). On this approach, 'knows' sometimes picks out a relation requiring that one meet standard s_1 with respect to a proposition, sometimes picks out a

This theory, as it stands, prompts at least two questions. First, what *are* epistemic standards and what does it take to meet them? Second, how do facts about the attributor's context determine an appropriate epistemic standard? DeRose addresses the first of these questions by modelling epistemic standards as *similarity spheres*. If S is one of the similarity spheres around a world w, say that E provides *safe*$_S$ evidence for P, at w, iff all of the E-worlds in S are P-worlds. If one believes P then the belief might be described as safe$_S$ iff it is based upon safe$_S$ evidence for P. According to DeRose a knowledge-of-P attribution is true only if the subject believes P and the belief is safe$_S$ where S is the similarity sphere that is appropriate given the attributor's context. The larger S is, the more difficult it will be for the subject's belief to be safe$_S$ and the more difficult it will be for the attribution to express a truth. For DeRose, the epistemic standards that can feature in the truth conditions of knowledge attributions, at a possible world w, correspond to the similarity spheres around w, and what it takes for one to meet a standard is for one to hold a belief that is safe throughout the corresponding similarity sphere.[12]

DeRose also supplies an answer to the second question above—one that exploits the idea of contextual *salience*. For DeRose, the similarity sphere that is appropriate in a given context will be the smallest similarity sphere to accommodate, as far as possible, the truth of all propositions that are salient to the attributor. The more propositions that are salient, and the more far-fetched they are, the larger the appropriate similarity sphere will be. While DeRose suggests that merely attending to a proposition may not be enough to make it salient in the relevant sense, his criteria for salience are, nevertheless quite lenient—the utterance of a knowledge-of-P attribution will suffice, for DeRose, to make both P and

relation requiring that one meet standard s_{17} with respect to a proposition, etc. Others regard the verb 'knows' as associated with a hidden argument place ranging over epistemic standards. On this view, an attribution of the form '... knows that __' has the underlying form '... knows that __ relative to standard s' where s is supplied by the context (see Ludlow 2005). Proponents of the former approach are committed, in effect, to a plurality of different knowledge relations, while proponents of the latter approach are committed to a single relation—albeit of greater *adicity* than has standardly been supposed. Other approaches are possible too. The precise linguistic implementation of contextualism won't assume much significance for present purposes.

[12] When DeRose first sets out this view in 'Solving the skeptical problem' (1995) he does not use the terminology of safety. In later work, he describes the theory in terms closer to these (DeRose 2004), as do certain commentators (Blome-Tillmann 2009).

~P salient. As such, for any knowledge-of-P attribution to be true, the subject's belief must be safe throughout the smallest sphere to permit ~P. If the subject believes P based upon E, then the attribution will be true only if, in the most similar possible worlds in which P is false, E is false—that is, only if the subject's belief is *sensitive* as well as safe. According to DeRose, whenever one makes a knowledge attribution, the standards for knowledge will automatically rise to the point at which a subject's belief must be sensitive in order to meet them. DeRose refers to this as the *Rule of Sensitivity* (DeRose 1995: s. 12).

One could develop a contextualist theory of justified belief attributions by substituting normic support for safety and world normalcy for world similarity throughout DeRose's theory. Staying close to the model provided by contextualism about knowledge attributions, contextualism about justified belief attributions would be the view that sentences of the form '. . . justifiably believes that __' are context sensitive in a way that owes specifically to the presence of the phrase 'justifiably believes that'. It would be the view that there are a range of epistemic standards, ranging from the lenient to the stringent, that can feature in the truth conditions of a justified belief attribution and that might be appropriate given the attributor's context. On this view, a justified belief-in-P attribution will be true only if the subject meets epistemic standard s with respect to P, where s is whatever standard is appropriate given the attributor's context. The more stringent this standard, the more difficult it will be for the subject to meet it and the more difficult it will be for the attribution to express a truth.[13]

If N is one of the normalcy spheres associated with a world w, say that E provides *normic support*$_N$ for P, at w, iff all of the E-worlds in N are P-worlds. If one believes P then the belief might be described as normically supported$_N$ iff it is based upon evidence that provides normic support$_N$

[13] Arguably, though, in thinking about contextualism about justified belief attributions, we shouldn't adhere too closely to the model provided by contextualism about knowledge attributions. It must be admitted that sentences of the form '. . . justifiably believes that __' don't roll off the tongue in quite the same way that sentences of the form '. . . knows that __' do. Indeed, such sentences may be quite uncommon outside of discussions in epistemology. This is not to say that ordinary speakers don't make judgements about epistemic justification—it's just that they may do so using different turns of phrase, such as 'It's reasonable for . . . to believe that __', '. . . has good reason to believe that __'. We may wish to broaden our definition of justified belief attributions to encompass sentences like these—and perhaps others besides.

for P. On the view that I envisage, a justified belief-in-P attribution is true only if the subject believes P and the belief is normically supported$_N$ where N is the normalcy sphere that is appropriate given the attributor's context. The larger N is, the more difficult it will be for the subject's belief to be normically supported$_N$ and the more difficult it will be for the attribution to express a truth. The epistemic standards that can feature in the truth conditions of justification attributions, at a possible world w, correspond to the normalcy spheres associated with w and what it takes for one to meet a standard is for one's belief to be normically supported throughout the corresponding normalcy sphere. On the view I envisage, the normalcy sphere that is appropriate in a given context will be the smallest normalcy sphere to accommodate, as far as possible, the truth of all propositions that are salient to the attributor. The more propositions that are salient, and the more abnormal they are, the larger the appropriate normalcy sphere will be.

The notion of salience invoked here could be regarded as the very same notion that is at play in DeRose's theory—though we may wish to specify, in addition, that the evidence on which a subject's belief is based will always count as salient when one attributes a justified belief to that subject. This will ensure that the normalcy sphere that features in the truth conditions of the attribution will always be one that permits the subject's evidence, irrespective of the context. Alternatively we could modify the theory to state that the appropriate normalcy sphere in a given context will be the smallest sphere to accommodate the truth of all propositions that are salient to the attributor, or the smallest sphere to accommodate the subject's evidence, whichever is larger.[14]

Contextualism about justified belief attributions is not a view that has received a great deal of attention from epistemologists—though the view has been defended by both Cohen (1999) and Wedgwood (2008). The theory proposed by Wedgwood is perhaps the most developed and systematic theory of this kind. Wedgwood sets up his theory within a

[14] If the appropriate normalcy sphere were determined not by what is salient to the attributor but, rather, by what is salient to the *subject*, we would arrive at a kind of 'interest relative invariantism' about justified belief attributions; a view on which the truth conditions of justified belief attributions do not vary according to the context of attribution, but whether these conditions are met by a subject may depend on what is salient to him, even if all other factors, such as the strength of his evidence, are held constant. Such a view would lie close to the interest relative threshold normic theory discussed in section 5.2.

broadly risk minimisation type framework—and the result is quite different from the sort of theory sketched here. It is not my intention, in any case, to defend the present DeRose-style contextualist theory of justified belief attributions—or any sort of contextualism about justified belief attributions. The present theory could, however, prove to be a useful supplement to a DeRose-style contextualism about knowledge attributions—and may be able to reproduce, in the case of justified belief attributions, many of its purported advantages.

One of the advertised advantages of DeRose's theory is it allows us to accommodate certain sceptical tendencies in our ordinary epistemic practices without becoming out and out sceptics. Suppose I'm at the zoo and I spy some black and white, striped, equine animals grazing in an enclosure marked 'Zebras' (Dretske 1970). At this point I'd be perfectly willing to assert, 'I know that those animals are zebras.' But if someone were to then raise the possibility that the animals could be mules cleverly disguised by the zoo authorities to look like zebras, I may become more apprehensive about making such an assertion. DeRose's theory offers an account of this. Once the disguised mule possibility is made salient to me, 'I know that those animals are zebras' would express a falsehood—my belief that the animals are zebras based on their black and white striped equine appearance will not be safe throughout a similarity sphere that includes disguised mule worlds. Out and out scepticism is resisted, however. Before the disguised mule possibility is made salient, 'I know that those animals are zebras' may express a truth—my belief may be safe throughout a smaller similarity sphere.

Should we wish it, the contextualist theory that I've described will generate the same predictions about a corresponding justified belief attribution. On this theory 'I justifiably believe that those animals are zebras' may express a truth until the disguised mule possibility is raised, at which point it will express a falsehood. My belief that the animals are zebras, based on their black and white striped equine appearance, will not be normically supported throughout a normalcy sphere that includes disguised mule worlds, but may be normically supported throughout a smaller normalcy sphere.[15] I won't pursue this further here.

[15] In section 3.2 I noted one 'mildly' sceptical consequence of any theory on which normic support is necessary for justification: one will lack justification for believing anything that entails a proposition about the outcome of a lottery, about which one has

6.4. Internalism and Externalism

I suggested in section 6.2 that whether a belief is justified is a more modally robust feature of that belief than whether it is safe or whether it counts as knowledge. Though it's not often expressed in quite this way, this idea would be common ground to many epistemologists. Some epistemologists, however, may be inclined to take this a step further. In section 6.2 I mentioned the absoluteness constraint requiring that all possible worlds are to share the same system of spheres. If all possible worlds were associated with the same system of normalcy spheres, then normic support relations would be *necessary* relations. Some epistemologists may be attracted by the idea that justificatory relations hold as a matter of necessity: if a body of evidence E provides justification for believing a proposition P, then it will be necessary that E provides justification for believing P and if E fails to provide justification for believing P, then it will be necessary that E fails to provide justification for believing P. This might be thought to chime with a kind of *internalism* about justification (see Pollock 1986: chap. 5, s. 4, Fumerton 1995; see also Comesaña 2010a).

There are, of course, a number of different ideas about epistemic justification that tend to be classed as 'internalist' and epistemologists who identify as internalists often understand the view in slightly different ways. If there is one thing that such epistemologists share it is a basic conviction that we must enjoy *some* sort of special, reflective access to facts about what we have justification for believing. If one's evidence provides justification for believing P then, for internalists, one should always be in a position to determine that one's evidence provides justification for

only statistical evidence. Suppose again that I'm due to meet a generally reliable friend for lunch tomorrow, but I know that she holds a ticket in a lottery that is about to be drawn and, should her ticket win, she will miss our lunch. We could imagine a contextualist theory on which 'I am justified in believing that I will meet my friend for lunch tomorrow' expresses a truth until such time as the possibility of her winning the lottery is raised at which point it will express a falsehood.

Whatever the merits of such a prediction, the contextualist theory under consideration could not deliver it. The possibility of my friend winning the lottery may not be one that I'm inclined to think about—but it would not be in any way *abnormal* and, as such, making it salient could not in and of itself affect any change in the truth conditions of justification attributions. Put differently, the smallest normalcy sphere to permit my evidence will *already* include worlds in which my evidence holds, my friend wins the lottery and misses our lunch—no expansion is necessary to find such worlds.

believing P just by carefully reflecting on the matter. The idea that we can always do this is, in a way, a comforting or empowering one—it seems to place us firmly in control of our epistemic destinies. The claim that justificatory relations are necessary relations may not guarantee this kind of special access all by itself—but it does remove one important obstacle to it. If justificatory relations are necessary relations, then whether one's evidence provides justification for believing a proposition will not be hostage to contingent, extra-evidential features of the world that are not accessible by reflection. Rather, whether one's evidence provides justification for believing a proposition will be determined solely by the content of that evidence and by the nature of the proposition in question. Henceforth, I shall refer to this view as internalism about justification and its denial as externalism about justification.[16]

Certain forms of reliabilism about justification provide the clearest examples of externalist theories. According to reliabilists, roughly speaking, a belief is justified just in case it was formed via a reliable process, or based upon evidence that reliably indicates its truth. Reliability, in turn, is usually spelled out in statistical terms: a piece of evidence reliably indicates the truth of a proposition iff believing a proposition of that kind on the basis of evidence of that kind would yield a high *proportion* of true beliefs in actual and similar counterfactual circumstances (see, for instance, Alston 1995: s. III, Comesaña 2009, 2010b: s. 4.4; see also Goldman 1979: s. II). Whether such a condition is met will, in general, be a contingent matter—and not something that could be determined by

[16] One could, of course, combine this kind of internalism with a broadly externalist account of evidence *possession* in which case the overall view that emerges may not make justification accessible in the way that the internalist desires. There are in effect two distinct points at which external conditions can enter into the justificatory story: first, the constitution of one's body of evidence might or might not depend upon external factors. The knowledge account of evidence is an example of an externalist theory of evidence, in this sense, while those who offer more restrictive accounts are often motivated by a kind of internalist instinct. Second, the bearing of one's body of evidence—facts about what propositions it provides justification for believing—might or might not depend upon broadly external factors. This is the issue that I take up in the body text. Pollock (1986: chap. 5) is naturally read as endorsing a view that is both internalist about evidence and evidential support. Comesaña (2010a) suggests a view that is internalist about evidence but externalist about evidential support (see also Alston 1988). Williamson (2000: chaps 9 and 10) can perhaps be read as defending the reverse view—one that is externalist about evidence and internalist about evidential support. It is more difficult to find thinkers who exemplify the final possible position—that which marries externalism about evidence possession and about evidential support—but this too is perfectly consistent.

reflection alone. One apparent advantage of reliabilism is that it secures a tight connection between justification and truth—justified beliefs are those that are highly *disposed* to be true, given the way they are formed or the evidence on which they are based. Internalism, on the other hand, would seem to lead to a greater separation between justification and truth.

The normic support condition on justification could also be described as reliabilist—and could be construed as an alternative way of spelling out the idea that one's evidence reliably indicates the truth of a proposition.[17] Unlike the statistical reliabilist condition though, the normic support condition does not force one's hand when it comes to the question of internalism. If we impose an absoluteness condition upon world normalcy, then every possible world will be assigned the same system of normalcy spheres. In this case, if one world is more normal than another, it is more normal *simpliciter*, more normal when viewed from *any* world (or from none). This leads to a kind of internalism *about normic support*—a view on which normic support relations are necessary relations.

According to internalism about normic support, if E normically supports P, then it will be necessary that E normically supports P and if E does not normically support P, then it will be necessary that E does not normically support P. Whether one's evidence provides normic support for believing a proposition will be determined solely by the content of the evidence and by the nature of the proposition in question. If normic support is both necessary and sufficient for justification then internalism

[17] While it may be somewhat natural to categorise my view as a kind of reliabilism, there are two broad reasons why I generally opt not to do so. The first reason is that, although the notion of normic reliability has, I suspect, just as strong a claim on our ordinary everyday 'reliability' talk as the notion of statistical reliability does, reliabilism tends to be very strongly associated with reliability of the latter sort. The second reason is that reliabilism often goes hand in hand with an ambition to carve out a place for epistemological notions within a purely naturalistic picture of the world—to find some way of reducing justification and knowledge to properties that are 'naturalistically respectable' (see, for instance, Zalabardo 2012: 3–4). The views I've defended here have not been put forward with any such ambition in mind, though I see no reason why they should frustrate it. One might think that statistical reliability is a more naturalistically respectable property than normic reliability is, since facts about the frequency of a certain occurrence do look like straightforwardly natural facts. On closer inspection, though, it's not at all clear that statistical reliabilist theories enjoy any 'naturalistic' advantage over normic reliabilist theories. In order to be remotely adequate, a statistical reliabilist condition must incorporate quantification over similar counterfactual scenarios—and it's not at all clear why this should be regarded as any more naturalistically respectable than the quantification over normal counterfactual scenarios that suffices to define normic reliability. I won't pursue this any further here.

about normic support is simply equivalent to internalism about justification. If normic support is merely necessary for justification, but not sufficient, then internalism about normic support won't inevitably lead to internalism about justification—but it will leave open the path to such a view.[18]

While internalism about normic support is simple to implement formally, we might wonder whether it really can be squared with any intuitive understanding of normalcy. The idea that standards of normalcy are somehow fixed, and don't vary from world to world, may seem strange at first. Certainly if normalcy were being understood in a *statistical* way, then this would clearly be wrong—as observed, facts about statistical frequencies are certainly contingent. But the notion that we are working with here is not that of statistical normalcy—it is that of *idealised* normalcy. As such, normic support relations will be necessary relations on the condition that all possible worlds end up being idealised in the same way—that the same set of worlds will serve as idealised models of any worlds whatsoever.

Suppose, once again, that I believe there is a barn before me on the grounds that there appears to be a barn before me. To simplify matters, suppose we accept a view on which perceptual experiences can represent high-level properties (Siewert 1998, Siegel 2006) and imagine that I'm simply undergoing a perceptual experience that represents the presence of a barn before me and I take this experience at face value. As we've seen, the presence of barn façades in the vicinity may prevent this belief from qualifying as safe or as knowledge, but would not prevent the belief from qualifying as normically supported or as justified. The most normal worlds in which I'm undergoing an experience as of a barn before me will be worlds in which there is a barn before me—the presence of barn façades in the vicinity will not change this. But would *anything* change this?

[18] It's interesting to note that Bird's (2007) theory of justification would seem to commit one to a view very close to this kind of internalism. According to Bird, recall, if a subject forms a belief, then that belief will count as justified iff there is a possible world in which the subject, with the same mental states, forms a corresponding belief and the belief qualifies as knowledge. Given the (5) modal axiom that what is possible is necessarily possible—or, equivalently, given that the world accessibility relation is Euclidean—it follows from this that if a subject holds a belief that counts as justified, then it is necessarily true that the subject's belief counts as justified, given his mental states as he formed it. Normic theories of justification, as noted, do not force us to go in either an internalist or an externalist direction.

One thing we might imagine is that the barn façades are not restricted to the local vicinity—perhaps, unbeknownst to me, there has been a far-reaching, worldwide conspiracy to erect such façades. We could imagine that the conspiracy has been so successful that the vast majority of the objects on Earth that would prompt an experience as of a barn are in fact façades and not the real thing. Certain forms of reliabilism about justification may predict at this point that my belief would no longer be justified—after all, believing that there's a barn before me upon undergoing a perceptual experience as of a barn before me would no longer be a method that would lead to a high proportion of true beliefs in actual and similar counterfactual circumstances. It seems to me, though, that this would not be the right prediction. Provided I am completely unaware of this conspiracy, a perceptual experience as of a barn before me would still provide justification for believing that there is a barn before me. And, in the idealised sense of 'normally', it would still be true that, if I have a perceptual experience as of a barn before me there would normally be a barn before me.

The actual world and the barn-façade conspiracy world would be idealised in the same way. A worldwide barn-façade conspiracy, no less than the presence of a few façades in the vicinity, is a feature of the world from which we should idealise away. In the most idealised worlds in which I'm having a perceptual experience as of a barn before me, there is no conspiracy and there is a barn before me.[19] These reflections don't

[19] A number of epistemologists have been attracted to the idea that, whenever one undergoes a perceptual experience, one always acquires prima facie justification for believing that things are as the experience represents them as being—if one has a perceptual experience as of P then one has prima facie justification for believing P (see, for instance, Audi 1993: 308, 366, Pryor 2000, 2004). This is sometimes termed the 'perceptual principle'. On the present view, the perceptual principle would require that a perceptual experience as of P always provides normic support for P. As I suggested in Chapter 3 n. 6, the claim that testimony always supplies normic support for its content is false on the grounds that people can testify to propositions, the obtaining of which would represent as great a departure from ideal normalcy as would a mistake or a lie. The claim that perception always supplies normic support for its content is beset by a similar difficulty. There will always be a trade-off between the abnormality of a given perceptual experience being a result of illusion, hallucination, deception, etc. and the abnormality of what the experience represents as being the case. Normic support will only flow on the condition that the former outstrips the latter. But the perceptual principle may yet survive, precisely because the representational potential of perception is *limited* in a way that the representational potential of testimony is not.

Generally speaking, the more constrained one takes the possible content of perceptual experiences to be, the less plausible it is that a perceptual experience could represent a state of affairs that implicates as great a departure from ideal normalcy as perceptual error, and the more secure the perceptual principle becomes. The less constrained one takes the

yet show, of course, that internalism about normic support is a viable position. Even if the actual world and the façade conspiracy world would be idealised in the same way, this doesn't show that *all* worlds should be. We can easily imagine worlds that are far more dissimilar than these two.

One of the best-known objections to reliabilism is the so-called 'new evil demon problem' first described by Cohen (1984). Consider a world inhabited by individuals much like ourselves, but presided over by a demon who delights in deceiving them. While these unfortunate individuals undergo perceptual experiences much like our own, these experiences are not produced by a functioning perceptual apparatus and reflect nothing about the actual environment—rather, they are manufactured by the demon. The inhabitants of the demon world are, of course, oblivious to their predicament—they take themselves, incorrectly, to be living in a world much like ours.

It is usually claimed that, in the demon world, one's perceptual experiences do not reliably indicate anything about one's environment. Certainly this would seem to be correct if reliability is being understood in a statistical way. If a demon world inhabitant has, say, an experience as of a red apple before him, then this would not *frequently* be accompanied by the presence of a real red apple. This much is effectively stipulated. If the demon worlder were to believe that there is a red apple before him, based on his perceptual experience, then statistical reliabilism would predict that this is an unjustified belief. Some reliabilists insist that this is the right prediction, urging that the demon worlder's belief has no connection to the truth—is no more disposed to be true than an arbitrary or superstitious belief would be at the actual world. And yet, to condemn this belief as unjustified strikes us overly harsh—even hypocritical, when combined with the observation that any of us would believe the *exact same thing* if placed in that very situation. There is a tension involved in epistemically criticising an individual while, at the same time, admitting that one could do no better if placed in their circumstances.

content of perceptual experiences to be, the more plausible it is that a perceptual experience could represent a state of affairs that implicates as great a departure from ideal normalcy as perceptual error and the more precarious the perceptual principle becomes. Either option, in any case, is perfectly compatible with internalism about normic support—with it being a matter of necessity which perceptual experiences are capable of supplying normic support for their contents.

If we understand reliability in the *normic* sense, rather than the statistical sense, then it cannot simply be *stipulated* that one's perceptual experiences, at the demon world, do not reliably indicate anything about one's environment. Whether this is so will depend upon questions about the standards of normalcy that the demon world imposes—and these questions cannot be settled by fiat. Perhaps the worlds which serve as simplified, idealised models of the actual world will serve still as simplified, idealised models of the demon world. Perhaps the actual world and the demon world are associated with the same normalcy spheres. If this is right, then normic support relations will come out the same in the two worlds and, in the normic sense, perception will be *equally reliable* at both. When the demon worlder believes that there is a red apple before him, this belief will be normically supported and may be justified. This belief *will* have a connection to the truth, more so than an arbitrary or superstitious belief, and every bit as much as a perceptual belief held at the actual world.

A contrasting thought is that the demon world and the actual world are so different that even their *idealisations* would be different—perhaps idealisation would take these worlds further apart rather than closer together. If this is right, then all bets are off—when the demon worlder believes that there is a red apple before him, the belief may not be normically supported or justified. I take no stand here on which of these is the right approach—or on whether internalism about normic support is ultimately a defensible view. My aim is merely to show what epistemic internalism will *amount to* if justification is understood in terms of normic support. If normic support is necessary for justification, then the question of whether justificatory relations are necessary will depend on the question of whether normic support relations are necessary which, in turn, will depend on questions about which possible worlds can serve as idealised models for others. If the view that all possible worlds are idealised in the same way can be defended, it will provide a means of being both an internalist *and* a reliabilist, at least of a sort.

In this chapter, I have shown how the normic support condition can be understood as a kind of modification of the safety condition that is widely thought to be necessary for knowledge. Nothing that I have said here amounts to a new *argument*, as such, for thinking that normic support is necessary for justification. Even those who think that safety is necessary for knowledge may still resist that conclusion. What I do hope to have shown is that combining a safety condition upon knowledge with a normic support condition upon justification can make for a simple and cohesive overall picture of these two norms—a picture on which knowledge and justification need not be so very different from one another.

On this picture, to describe a belief as knowledge or as justified conveys that the belief enjoys a kind of protection against the possibility of error. It's just that the kinds of error possibilities at issue are slightly different for each description. If a belief is to count as knowledge, it must be protected from error possibilities that could have easily obtained—error possibilities that would not represent a gratuitous departure from the world as it actually is. If a belief is to count as justified, it must be protected from error possibilities that could have normally obtained—error possibilities that would not represent a gratuitous departure from the world as it would ideally be. The notions of knowledge and justification are used to keep track of vulnerability to error possibilities of each sort.

7

Similar Worlds, Normal Worlds

In this chapter I will explore the safe evidence and normic support relations through the lens of sphere models for variably strict conditional logic. The formal features of safety and of normic support will be reflected in the validity and invalidity of various inference patterns for variably strict conditional operators. I will contrast the formal structure of normic support with the formal structure of a relation of *probabilistic support* and argue that the former seems a much better fit with the formal structure that justification might be thought to possess. This chapter, and Chapters 8 and 9, are more technical than the chapters that have come before, and include proofs of a number of results. In some cases, these proofs offer insight into why the result in question holds. In other cases, the result may be rather unsurprising, and the proof is provided merely for the sake of thoroughness. Whatever the case, the proofs can, for the most part, be skipped over without any loss of comprehension.

7.1. The Formal Structure of Safety

A *sphere model* consists of a set of worlds W and a function \$ taking members of W to sets of subsets of W. That is, $\$: W \rightarrow \wp(\wp(W))$. We term $\$_w$ the *system of spheres* associated with w. Sphere systems are required to be non-empty: for any $w \in W$, $\$_w \neq \emptyset$. Sphere systems are required to be nested: for any $w \in W$ and $S, S' \in \$_w$ either $S \subseteq S'$ or $S' \subseteq S$. Finally, sphere systems are required to be closed under unions and non-empty intersections: for any $w \in W$ if $\Sigma \subseteq \$_w$ then $\cup\Sigma \in \$_w$ and if, in addition, $\Sigma \neq \emptyset$

then $\cap\Sigma \in \$_{w}$.[1] Notice that closure under unions guarantees, first, that the empty set will automatically count as one of the spheres associated with a world w and, second, that there will be a largest sphere associated with a world w—namely $\cup\$_w$. Informally, we may think of W as the set of metaphysically possible worlds or as the set of (centered) epistemically possible worlds and we think of $\$_w$ as carrying information about the comparative similarity of worlds to w. Lewis, of course, devised sphere models in order to provide truth conditions for the sentences of a formal language. Here we abstract away from the semantic features of such models and deal directly with propositions, which might be modelled as subsets of W. For a proposition P we say that P is true at a world w or that w is a P-world just in case $w \in P$. In using this model we needn't suppose, as some do, that propositions literally *are* sets of possible worlds (Lewis 1973: s. 2.2, Lewis 1979, Stalnaker 1976, 1984). It will be enough that the notions being modelled are not sensitive to differences between propositions that share the same possible worlds truth conditions.

The extensional propositional operators $\wedge$, $\vee$, $\supset$, $\sim$ can all be defined in the usual way. Say that a sphere S permits a proposition P just in case there is a world $w \in S$ at which P is true. We define a further propositional operator $\Box\!\!\rightarrow$ as follows:

> E $\Box\!\!\rightarrow$ P is true at $w \in W$ iff either (i) there is a sphere $S \in \$_w$ such that S is E-permitting and every E-world in S is a P-world or (ii) there is no sphere $S \in \$_w$ such that S is E-permitting.

Informally, we think of E $\Box\!\!\rightarrow$ P as the proposition that E provides safe evidence for P.

Recall that $\cup\$_w$ is guaranteed to be the largest sphere in $\$_w$. With this in mind, we define two further propositional operators as follows:

> $\Box$P is true at $w \in W$ iff every world in $\cup\$_w$ is a P-world.
> E $\rightarrow$ P is true at $w \in W$ iff every E-world in $\cup\$_w$ is a P-world.[2]

[1] Any finite system of nested spheres (that includes the empty set) will automatically satisfy closure under unions and non-empty intersections. These conditions are needed in the case of an infinite system of spheres which could, otherwise, contain infinite sequences of larger and larger or smaller and smaller spheres without including their smallest upper, or largest lower, bound.

[2] $\Box$ and $\rightarrow$ can be defined in terms of $\Box\!\!\rightarrow$ as follows:

$$\Box\, P \equiv (\sim P \;\Box\!\!\rightarrow P)$$
$$(P \rightarrow Q) \equiv ((P \wedge \sim Q) \;\Box\!\!\rightarrow Q)$$

The largest sphere in $\$_w$ can be understood as comprising all of those possible worlds that can be assessed for their similarity to w. As briefly mentioned in section 6.1, it's natural to think that this is simply the class of worlds that are possible from the perspective of w—if a world is possible from the perspective of w, then it can be assessed for its similarity to w and if it can be assessed for its similarity to w, then it is possible from the perspective of w. If W is understood as the set of metaphysically possible worlds, then $\Box P$ could be understood as the proposition that P is metaphysically necessary and $E \rightarrow P$ as the proposition that E implies P as a matter of metaphysical necessity. If W is understood as the set of epistemically possible worlds, then $\Box P$ could be understood as the proposition that P is epistemically necessary or 'a priori' and $E \rightarrow P$ as the proposition that E implies P as a matter of epistemic necessity or that E implies P a priori.[3] I will make some use of these operators in due course.

As well as the compulsory constraints of non-emptiness, nesting, and closure under unions and (non-empty) intersections, Lewis also considers a number of optional constraints that might be imposed upon systems of spheres. According to the *centering* constraint, the smallest non-empty sphere associated with a world will always be a singleton set containing that world—for any world $w \in W$, $\{w\} \in \$_w$. According to the *weak centering* constraint, there is always a smallest non-empty similarity sphere associated with any world and that world is guaranteed to be a member of it—for any world $w \in W$, there is a sphere $S \in \$_w$ such that $S \neq \emptyset$, for any $S' \in \$_w$ if $S' \subset S$ then $S' = \emptyset$ and $w \in S$. Weak centering is, of course, entailed by centering but the converse does not hold.

[3] Given the constraints that have been imposed thus far, $\Box$ will be governed by the modal logic **K**. By imposing further constraints upon the composition of similarity sphere systems we can strengthen this logic in various ways. Suppose we stipulate that, for any world w, there is some $S \in \$_w$ such that $w \in S$. This will correspond, in the standard accessibility semantics, to a reflexivity constraint and will serve to validate all substitution instances of the axiom schema (T): $\Box P \supset P$. This will be guaranteed by a centering or a weak centering constraint upon systems of similarity spheres but is, in fact, a good deal weaker than either (Lewis terms this constraint *total reflexivity*). Suppose we stipulate that, for any worlds w_1 and w_2, if w_1 is a member of some sphere $S \in \$_{w2}$ then w_2 is a member of some sphere $S' \in \$_{w1}$. This will correspond to a symmetry constraint in the standard accessibility semantics and validate all substitution instances of the (B) axiom schema: $P \supset \Box\Diamond P$. Finally, suppose we stipulate that, for any worlds w_1, w_2, and w_3, if w_2 is a member of some sphere $S \in \$_{w1}$ and w_3 is a member of some sphere $S' \in \$_{w2}$ then w_3 is a member of some sphere $S'' \in \$_{w1}$. This will correspond to the transitivity constraint in the accessibility semantics and will validate all substitution instances of the (4) axiom schema: $\Box P \supset \Box\Box P$.

The centering constraint guarantees the validity of the following inference pattern, sometimes called *conjunctive sufficiency*:

$$\frac{\begin{array}{l}E\\P\end{array}}{E \;\Box\!\!\rightarrow P}$$

Proof By centering, for any world $w \in W$, $\{w\} \in \$_w$. As such, if E and P are both true at w, then there is guaranteed to be a sphere in $\$_w$ which permits E and in which all E-worlds are P-worlds, in which case E □→ P is true at w as required. **QED**

I noted in section 6.1 that the centering constraint makes the operator □→ ill suited for capturing the safe evidence relation. Suppose I possess a body of evidence E. Assuming that evidence must consist of truths it follows, by conjunctive sufficiency, that E □→ P for any true proposition P. It is not the case, though, that I possess safe evidence for every true proposition or that all true beliefs count as safe. If we replace centering with weak centering, then conjunctive sufficiency will no longer be valid, making □→ a better formal fit with the safe evidence relation.

With weak centering in place of centering, it will no longer be the case that all true beliefs are safe. It will still be the case, however, that all safe beliefs are true. This is guaranteed by the fact that *modus ponens* will still be a valid inference pattern for □→:

$$\frac{\begin{array}{l}E\\E \;\Box\!\!\rightarrow P\end{array}}{P}$$

Proof Suppose that E and E □→ P are both true at a world $w \in W$. Let S be the smallest non-empty sphere in $\$_w$. By weak centering, $w \in S$. Since E is true at w, S is E-permitting. Since E □→ P is true at w, all E-worlds in S are P-worlds. It follows that w must be a P-world. **QED**

Given weak centering as our only optional constraint, the safe evidence relation will be governed by the logic Lewis terms **VW** (Lewis 1973: chap. 6).[4]

[4] On the axiomatisation that Lewis provides, **VW** has, as axioms, all truth functional tautologies, definitions of □ and → in terms of □→ and all instances of the following schemas:

(1) P □→ P

(2) (~P □→ P) ⊃ (Q □→ P)

7.2. The Formal Structure of Normic Support

We now augment our sphere models by adding a further function ₦ taking members of W to sets of subsets of W. That is, ₦: W $\rightarrow \wp(\wp(W))$. Every world in W will now be associated with *two* systems of spheres; $\$_w$ and $₦_w$. Formally, ₦ behaves very much like \$. That is, ₦ is required to satisfy non-emptiness, nesting, and closure under unions and non-empty intersections. The only formal difference, initially at any rate, is that ₦ is not required to satisfy a weak centering constraint. Informally, we understand $₦_w$ as carrying information about the comparative normalcy of possible worlds from the perspective of w.

We define a further propositional operator ■→ as follows:

> E ■→ P is true at $w \in W$ iff either (i) there is a sphere $N \in ₦_w$ such that N is E-permitting and every E-world in N is a P-world or (ii) there is no sphere $N \in ₦_w$ such that N is E-permitting.

Informally, we think of E ■→ P as the proposition that E normically supports P.

Closure under unions guarantees that, for any world $w \in W$, there will be a largest sphere in $₦_w$, namely $\cup ₦_w$. This sphere can be understood as comprising all of the possible worlds that can be assessed for their normalcy from the perspective of w. It is, once again, quite natural to regard this as simply being the set of worlds that are possible from the perspective of w. I will assume, at any rate, that for any world w, $\cup ₦_w = \cup \$_w$. In this case, the operators □ and → could have equally been defined in terms of ₦ or the operator ■→.

In the absence of the centering and weak centering constraints, there is no guarantee that $₦_w$ contain any non-empty spheres at all. To provide this guarantee, we can introduce a constraint that Lewis termed

(3) $(P \,\Box\!\!\rightarrow \sim Q) \vee (((P \wedge Q) \,\Box\!\!\rightarrow R) \equiv (P \,\Box\!\!\rightarrow (Q \supset R)))$
(4) $(P \,\Box\!\!\rightarrow Q) \supset (P \supset Q)$

There are three inference rules—modus ponens, the substitution of logical equivalents, and the following, which Lewis terms *Deduction With Conditionals*:

For any $n \geq 1$

$$\frac{\vdash (P_1 \wedge \ldots \wedge P_n) \supset Q}{\vdash ((E \,\Box\!\!\rightarrow P_1) \wedge \ldots \wedge (E \,\Box\!\!\rightarrow P_n)) \supset (E \,\Box\!\!\rightarrow Q)}$$

non-triviality: for any world $w \in W$, $Ꞥ_w \neq \{\emptyset\}$. By stipulating that $\cup Ꞥ_w = \cup \$_w$, we arrive at this in a slightly roundabout way. A further constraint is also smuggled in via this route: since $\cup \$_w$ must, by weak centering, contain w, $\cup Ꞥ_w$ must also contain w if the two are identical. Lewis terms this constraint *total reflexivity*.

Modus ponens turns out to be an invalid inference pattern for ■→:

$$\frac{\begin{array}{c}E\\ E \blacksquare\!\!\rightarrow P\end{array}}{P}$$

Proof Consider two worlds w_1 and $w_2 \in W$. Suppose that E is true at both worlds while P is false at w_1 and true at w_2. Suppose, finally, that $\{w_2\} \in Ꞥ_{w1}$. If we had a weak centering constraint in place then this, of course, could not happen—w_1 would have to be a member of any non-empty sphere in $Ꞥ_{w1}$. Without such a constraint, though, there is nothing to prevent such a stipulation. Since $\{w_2\}$ is E-permitting and all the E-worlds in $\{w_2\}$ are P-worlds it follows that E ■→ P is true at w_1 as required. **QED**

This failure of modus ponens is quite apposite if ■→ is to represent the relation of normic support. This reflects the fact that normic support is *defeasible*—one can possess evidence that supports a false proposition. Normically supported beliefs, unlike safe beliefs, do not have to be true. Modus ponens, as noted, is a valid pattern for □→—its validity being characteristic of the weak centering constraint.

Another notable inference pattern that is invalid for ■→ is the pattern sometimes termed *strengthening the antecedent*:

$$\frac{E \blacksquare\!\!\rightarrow P}{(E \wedge F) \blacksquare\!\!\rightarrow P}$$

Proof Consider three worlds w_1, w_2, and $w_3 \in W$. Suppose E, F, and P are all false at w_1. Suppose that E and P are true and F is false at w_2 and that E and F are true and P is false at w_3. Suppose, finally, that $\{w_1, w_2\} \in Ꞥ_{w1}$ and $\{w_1, w_2, w_3\} \in Ꞥ_{w1}$. Since $\{w_1, w_2\}$ is E-permitting and all of the E-worlds in $\{w_1, w_2\}$ are P-worlds, it follows that E ■→ P is true at w_1. Since $\{w_1, w_2, w_3\}$ is the smallest $E \wedge F$-permitting sphere in $Ꞥ_{w1}$ and it contains a world at which $E \wedge F$ is true and P is false, it follows that $(E \wedge F)$ ■→ P is false at w_1 as required. **QED**

This provides us with another way of thinking about the defeasibility of normic support—just because a body of evidence normically supports a proposition, it does not follow that the proposition will be normically supported by an expanded or enriched body of evidence. This property is sometimes termed *non-monotonicity*. If E normically supports P and $E \wedge D$ does not normically support P then we might say that D *defeats* the normic support relation between E and P. Defeat can occur when the smallest $E \wedge D$-permitting sphere in ₦_w is larger than the smallest E-permitting sphere in ₦_w. The acquisition of new evidence can, in effect, force us further away from idealised normalcy—force us to evaluate the bearing of our evidence relative to an expanded set of worlds. D will defeat the normic support relation between E and P if it takes us to an expanded set of worlds in which the modal relationship between E and P breaks down.

Strengthening the antecedent is also an invalid pattern for $\Box\!\!\rightarrow$—as can be easily checked, the countermodel sketched above does satisfy the weak centering constraint. This suggests that these two ways of thinking about defeasibility can come apart from one another—the safe evidence relation is not defeasible in the sense that one can possess safe evidence for a falsehood, but it is defeasible in the sense of being non-monotonic. Significantly, though, while $(E \wedge F) \Box\!\!\rightarrow P$ will not follow from $E \Box\!\!\rightarrow P$ alone, it will follow if we add E and F as additional premises. This, then, is a valid inference pattern:

$$\begin{array}{l} E \\ F \\ E \Box\!\!\rightarrow P \\ \hline (E \wedge F) \Box\!\!\rightarrow P \end{array}$$

Proof Suppose that E and F and $E \Box\!\!\rightarrow P$ are true at a world $w \in W$. Let S be the smallest non-empty sphere in $\$_w$. By weak centering, $w \in S$. Since E is true at w, S is E-permitting. Since $E \Box\!\!\rightarrow P$ is true at w, all E-worlds in S are P-worlds. Since F is true at w, S is $E \wedge F$-permitting. Since all of the E-worlds in S are P-worlds it follows that all of the $E \wedge F$-worlds in S must be P-worlds in which case $(E \wedge F) \Box\!\!\rightarrow P$ is true at w as required. **QED**

Suppose one believes P based on E at world w and that E is safe evidence for P at w. While there may be *possible* propositions that could defeat the safe evidence relationship between E and P, no proposition that is true at w could have this effect. That is, any expanded or enriched body of evidence obtainable at w would continue to supply safe evidence for P. Safe evidence, we might say, is *hypothetically*, though not *actually*, defeasible.[5]

[5] The distinction between hypothetical and actual defeat arises in connection with so-called 'defeasibility theories' of knowledge. The simplest such theory seeks to analyse knowledge as justified, true belief where one's justification is required to be invulnerable to defeaters (Pappas and Swain 1978: 28). Requiring an invulnerability to all *conceivable* defeaters, however, would effectively reduce this theory to a kind of infallibilism about knowledge. By requiring an invulnerability only to *actual* defeaters, one arrives at a theory that is more lax than infallibilism, and yet more demanding than the traditional justified, true belief theory—and more demanding, or so it is hoped, in precisely the ways needed to circumvent the Gettier problem.

The idea that safety and normic support are necessary for knowledge and justification respectively may turn out to be an unexpected source of support for this theory—for it may well be that safety and normic support stand in the very relation which this defeasibility theory envisages for knowledge and justification. Given a certain formal constraint it can be proved that if E provides normic support for P that is invulnerable to actual defeaters, then E must provide safe evidence for P. The constraint required is one that places limits on the *interaction* between $\$$ and ₦: for any three worlds w_1, w_2, and $w_3 \in W$, if w_2 is a member of a sphere N in ₦$_{w1}$ and w_3 is a member of the smallest non-empty sphere S in $\$_{w2}$ then w_3 is also a member of N. What this amounts to, informally, is that if w_2 estimates w_3 to be very similar to itself, then a third world w_1 cannot estimate w_3 to be any less normal than w_2.

With this constraint in place the proof proceeds as follows:

> *Proof* Assume, for reductio, that $(E \wedge {\sim}(E \Box\!\!\rightarrow P)) \blacksquare\!\!\rightarrow P$ is true at a world $w_1 \in W$. It follows that there must be a sphere in ₦$_{w1}$ that permits $E \wedge {\sim}(E \Box\!\!\rightarrow P)$ and in which all of the $E \wedge {\sim}(E \Box\!\!\rightarrow P)$-worlds are P-worlds—call this sphere N. Since N is $E \wedge {\sim}(E \Box\!\!\rightarrow P)$-permitting, there must be some world in N at which it holds—call this world w_2. Since ${\sim}(E \Box\!\!\rightarrow P)$ is true at w_2 it follows that there is a world in the smallest non-empty sphere in $\$_{w2}$ in which E is true and P is false—call this world w_3. Since E is true and P is false at w_3 it further follows, given the validity of modus ponens for $\Box\!\!\rightarrow$, that ${\sim}(E \Box\!\!\rightarrow P)$ is true at w_3. By the constraint noted above, $w_3 \in N$. But w_3 is a world in which $E \wedge {\sim}(E \Box\!\!\rightarrow P)$ is true and P is false, in which case not all of the $((E \wedge {\sim}(E \Box\!\!\rightarrow P))$-worlds in N are P-worlds, contrary to assumption. By reductio, we have it that ${\sim}((E \wedge {\sim}(E \Box\!\!\rightarrow P)) \blacksquare\!\!\rightarrow P)$ is true at w_1. Suppose now that E provides normic support for P at w_1—that $E \blacksquare\!\!\rightarrow P$ is true at w_1. If this normic support relation is invulnerable to actual defeat, then there are no propositions X such that X is true at w_1 and ${\sim}((E \wedge X) \blacksquare\!\!\rightarrow P)$. But it follows immediately that ${\sim}(E \Box\!\!\rightarrow P)$ cannot be true at w_1, which is just to say that $E \Box\!\!\rightarrow P$ *is* true at w_1 and E provides safe evidence for P at w_1. **QED**

The formal constraint required to drive this proof may well be problematic—but I won't examine its credentials further here. In any case, I don't mean to endorse the simple defeasibility theory of knowledge, though I do think that this theory has a lot more going for it than is commonly appreciated.

The corresponding inference pattern is not, however, valid for $\blacksquare\!\!\rightarrow$:

$$\begin{array}{l} E \\ F \\ E \blacksquare\!\!\rightarrow P \\ \hline (E \wedge F) \blacksquare\!\!\rightarrow P \end{array}$$

Proof Consider two worlds w_1 and $w_2 \in W$. Suppose that E and F are true and P is false at w_1 and that E and P are true and F is false at w_2. Suppose, finally, that $\{w_2\} \in \mathbb{N}_{w1}$ and $\{w_1, w_2\} \in \mathbb{N}_{w1}$. Since $\{w_2\}$ is E-permitting and all of the E-worlds in $\{w_2\}$ are P-worlds, it follows that $E \blacksquare\!\!\rightarrow P$ is true at w_1. Since $\{w_1, w_2\}$ is the smallest $E \wedge F$-permitting sphere in $\mathbb{N}_{w1}$ and it contains a world at which $E \wedge F$ is true and P is false, it follows that $(E \wedge F) \blacksquare\!\!\rightarrow P$ is false at w_1 as required. **QED**

Suppose that one believes P based on E at w and that E normically supports P at w. It is perfectly possible that there may be further propositions, true at w, that serve to defeat this normic support relationship. It is perfectly possible that there are expanded or enriched bodies of evidence, obtainable at w, that would not normically support P. Normic support is actually (as well as hypothetically) defeasible.

There are nevertheless certain formal constraints on when a proposition might serve to defeat a normic support relationship. One of these was described in section 4.3: if a piece of evidence normically supports a proposition, then it must also normically support the negation of any proposition that would serve to defeat this normic support. Put slightly differently, only a proposition whose negation is normically supported by a piece of evidence can serve to defeat any of the normic support relations that emanate from that evidence. This property is sometimes termed *rational monotonicity* and is reflected in the validity of the following inference pattern:

$$\begin{array}{l} E \blacksquare\!\!\rightarrow P \\ \sim(E \blacksquare\!\!\rightarrow \sim Q) \\ \hline (E \wedge Q) \blacksquare\!\!\rightarrow P \end{array}$$

Proof Suppose $E \blacksquare\!\!\rightarrow P$ and $\sim(E \blacksquare\!\!\rightarrow \sim Q)$ are both true at a world $w \in W$. Since $E \blacksquare\!\!\rightarrow P$ is true at w either (i) there are no E-permitting spheres in $\mathbb{N}_w$ or (ii) there is an E-permitting sphere in $\mathbb{N}_w$ in which all of the E-worlds are P-worlds. Suppose there are no E-permitting spheres in

₦_w. It follows at once that there are no E ∧ Q-permitting spheres in ₦_w in which case (E ∧ Q) ■→ P is true at w as required. Suppose there is an E-permitting sphere in ₦_w in which all of the E-worlds are P-worlds—call it N. Since ~(E ■→ ~Q) is true at w, every E-permitting sphere in ₦_w must also permit E ∧ Q. It follows that N must permit E ∧ Q. Since all of the E-worlds in N are P-worlds, all of the E ∧ Q-worlds in N must be P-worlds in which case (E ∧ Q) ■→ P is true at w as required. **QED**

If E normically supports P and does not normically support ~Q then Q cannot defeat the normic support relation between E and P.

Given the validity of the above inference pattern, it's easy to see that the following pattern will also be valid:

$$\begin{array}{c} E \blacksquare\!\!\rightarrow P \\ \underline{E \blacksquare\!\!\rightarrow Q} \\ (E \wedge Q) \blacksquare\!\!\rightarrow P \end{array}$$

Proof Suppose that E ■→ P and E ■→ Q are true at a world w ∈ W. Either (i) there are no E-permitting spheres in ₦_w or (ii) there is an E-permitting sphere in ₦_w in which all of the E-worlds are P-worlds and an E-permitting sphere in ₦_w in which all of the E-worlds are Q-worlds. Suppose there are no E-permitting spheres in ₦_w. It follows right away that (E ∧ Q) ■→ P is true at w as required. Suppose there is an E-permitting sphere in ₦_w in which all of the E-worlds are P-worlds—call it N—and an E-permitting sphere in ₦_w in which all of the E-worlds are Q-worlds—call it N′. By nesting, either N ⊆ N′ or N′ ⊆ N. In the former case, all of the E-worlds in N must be Q-worlds as well as P-worlds. In the latter case, all of the E-worlds in N′ must be P-worlds as well as Q-worlds. In either case there exists a sphere in ₦_w which permits E ∧ Q and in which all of the E ∧ Q-worlds are P-worlds in which case (E ∧ Q) ■→ P is true at w as required. **QED**

If E normically supports P and normically supports Q then Q cannot defeat the normic support relation between E and P. This is the principle of *cautious monotonicity* for normic support.

If E normically supports P then, by cautious monotonicity, P is prevented from defeating any of the normic support relations extending from E. Significantly, though, the reverse does not hold—E is *not* prevented from defeating normic support relations that extend from P. Defeat of this kind can, in fact, occur. The proposition that an object is a cleverly constructed paper mache barn façade normically supports the proposition that the object looks like a barn which, in turn, normically supports the proposition that the object is a barn—but the first proposition clearly defeats the normic support relation between the second and the third. This leads us to the invalidity of another notable inference pattern—*transitivity*:

$$\frac{\begin{array}{c} E \blacksquare\!\!\rightarrow P \\ P \blacksquare\!\!\rightarrow Q \end{array}}{E \blacksquare\!\!\rightarrow Q}$$

Proof Consider two worlds w_1 and $w_2 \in W$. Suppose that P and Q are true and E is false at w_1 and E and P are true and Q is false at w_2. Suppose, finally, that $\{w_1\} \in \text{₦}_{w1}$ and $\{w_1, w_2\} \in \text{₦}_{w1}$. Since $\{w_1\}$ is P-permitting and all of the P-worlds in $\{w_1\}$ are Q-worlds, it follows that P $\blacksquare\!\!\rightarrow$ Q is true at w_1. Since $\{w_1, w_2\}$ is E-permitting, and all of the E-worlds in $\{w_1, w_2\}$ are P-worlds, it follows that E $\blacksquare\!\!\rightarrow$ P is true at w_1. Since $\{w_1, w_2\}$ is the smallest E-permitting sphere in ₦_{w1} and it contains a world at which E is true and Q is false, it follows that E $\blacksquare\!\!\rightarrow$ Q is false at w_1 as required. **QED**

If E normically supports P and P normically supports Q it does not automatically follow that E normically supports Q. Reasoning along chains of normic support can lead us to conclusions not normically supported by our initial evidence. This result is somewhat less surprising, though, in light of the observation made above—that E normically supporting P and P normically supporting Q are quite compatible with E *defeating* the normic support relation between P and Q. And the three propositions used above to illustrate this provide a very clear counterexample to transitivity. It turns out, in fact, that this is the only kind of situation in which transitivity can fail—which points us towards a revised inference pattern that does prove valid:

$$\frac{\begin{array}{c} E \blacksquare\!\!\rightarrow P \\ (E \wedge P) \blacksquare\!\!\rightarrow Q \end{array}}{E \blacksquare\!\!\rightarrow Q}$$

Proof Suppose that E $\blacksquare\!\!\rightarrow$ P is true at a world $w \in W$. Either (i) there are no E-permitting spheres in ₦_w or (ii) there is an E-permitting sphere in ₦_w in which all of the E-worlds are P-worlds. Suppose there are no E-permitting spheres in ₦_w. It follows right away that E $\blacksquare\!\!\rightarrow$ Q is true at w as required. Suppose there is an E-permitting sphere in ₦_w in which all of the E-worlds are P-worlds—call it N. Suppose that $(E \wedge P) \blacksquare\!\!\rightarrow Q$ is true at w. It follows that there is an $E \wedge P$-permitting sphere in $\mathbb{N}_w$ in which all of the $E \wedge P$-worlds are Q-worlds—call it N′. By nesting, either $N \subseteq N'$ or $N' \subseteq N$. Since all of the E-worlds in N are $E \wedge P$-worlds, and all the $E \wedge P$-worlds in N′ are Q-worlds, if $N \subseteq N'$ then all E-worlds in N must be Q-worlds. Since all of the $E \wedge P$-worlds in N′ are Q-worlds and all the E-worlds in N are $E \wedge P$-worlds, if $N' \subseteq N$ then all E-worlds in N′ must be Q-worlds. In either case $E \blacksquare\!\!\rightarrow Q$ is true at w. **QED**

Given this inference pattern, if E normically supports P and E and P *together* normically support Q, then E alone normically supports Q. The validity of this pattern reflects the fact that normic support is *cumulative*—one can always add to one's body of evidence any propositions normically supported by that evidence in the service of drawing further normically supported conclusions. This is the principle of *cumulative transitivity* for normic support.

Another valid inference pattern that closely resembles the invalid transitivity pattern is *weakening the consequent*:

$$\frac{\begin{array}{c} E \blacksquare\!\!\rightarrow P \\ P \rightarrow Q \end{array}}{E \blacksquare\!\!\rightarrow Q}$$

Proof Suppose that $E \blacksquare\!\!\rightarrow P$ and $P \rightarrow Q$ are true at a world $w \in W$. Since $E \blacksquare\!\!\rightarrow P$ is true at w either (i) there are no E-permitting spheres in $\mathbb{N}_w$ or (ii) there is an E-permitting sphere in $\mathbb{N}_w$ in which all of the E-worlds are P-worlds. Suppose there are no E-permitting spheres in $\mathbb{N}_w$. It follows right away that $E \blacksquare\!\!\rightarrow Q$ is true at w as required. Suppose there is an E-permitting sphere in $\mathbb{N}_w$ in which all of the E-worlds are P-worlds—call it N. Since $P \rightarrow Q$ is true at w all of the P-worlds in $\cup\mathbb{N}_w$ are Q-worlds. Since N is a subset of $\cup\mathbb{N}_w$, all of the P-worlds in N are Q-worlds. It follows that all of the E-worlds in N are Q-worlds in which case $E \blacksquare\!\!\rightarrow Q$ is true at w as required. **QED**

The validity of weakening the consequent simply reflects the fact that normic support is closed under single premise deductive consequence—if evidence normically supports a proposition then it normically supports all of its deductive consequences.

As discussed in previous chapters, normic support is also closed under *multiple premise* deductive consequence—if evidence normically supports each of a set of propositions then it will support their joint deductive consequences. This is reflected in the additional fact that *agglomeration* is a valid pattern for the normic conditional:

$$\frac{\begin{array}{l} E \blacksquare\!\!\rightarrow P \\ E \blacksquare\!\!\rightarrow Q \end{array}}{E \blacksquare\!\!\rightarrow (P \wedge Q)}$$

Proof Suppose that $E \blacksquare\!\!\rightarrow P$ and $E \blacksquare\!\!\rightarrow Q$ are true at a world $w \in W$. Either (i) there are no E-permitting spheres in $\aleph_w$ or (ii) there is an E-permitting sphere in $\aleph_w$ in which all of the E-worlds are P-worlds and an E-permitting sphere in $\aleph_w$ in which all of the E-worlds are Q-worlds. Suppose there are no E-permitting spheres in $\aleph_w$. It follows right away that $E \blacksquare\!\!\rightarrow (P \wedge Q)$ is true at w as required. Suppose there is an E-permitting sphere in $\aleph_w$ in which all of the E-worlds are P-worlds—call it N—and an E-permitting sphere in $\aleph_w$ in which all of the E-worlds are Q-worlds—call it N′. By nesting, either $N \subseteq N'$ or $N' \subseteq N$. In the former case, all of the E-worlds in N must be P-worlds and Q-worlds. In the latter case, all of the E-worlds in N′ must be P-worlds and Q-worlds. In either case $E \blacksquare\!\!\rightarrow (P \wedge Q)$ is true at w. **QED**

Using agglomeration and weakening the consequent we can easily prove a principle reflecting full multiple premise closure: for any set of propositions $P_1 \ldots P_n$, if we have it that $E \blacksquare\!\!\rightarrow P_1 \ldots E \blacksquare\!\!\rightarrow P_n$ and that $(P_1 \wedge \ldots \wedge P_n) \rightarrow Q$, it follows that $E \blacksquare\!\!\rightarrow Q$.

The flipside of agglomeration is an inference pattern sometimes called *amalgamation.*

$$\frac{\begin{array}{l} E \blacksquare\!\!\rightarrow P \\ F \blacksquare\!\!\rightarrow P \end{array}}{(E \vee F) \blacksquare\!\!\rightarrow P}$$

This inference is also valid.

Proof Suppose that $E \blacksquare\!\!\rightarrow P$ and $F \blacksquare\!\!\rightarrow P$ are true at a world $w \in W$. There are four possibilities: (i) There are no E-permitting spheres in $\aleph_w$ and no F-permitting spheres in $\aleph_w$. (ii) There are no E-permitting spheres in $\aleph_w$ and there is an F-permitting sphere in $\aleph_w$ in which all of the F-worlds are P-worlds. (iii) There are no F-permitting spheres in $\aleph_w$ and there is an E-permitting sphere in $\aleph_w$ in which all of the E-worlds are P-worlds. (iv) There is an F-permitting sphere in $\aleph_w$ in which all of the F-worlds are P-worlds and there is an E-permitting sphere in $\aleph_w$ in which all of the

E-worlds are P-worlds. If (i) there are no $E \vee F$-permitting spheres in $\mathbb{N}_w$ in which case $(E \vee F) \blacksquare\!\!\rightarrow P$ is true at w as required. If (ii), let N be the F-permitting sphere in $\mathbb{N}_w$ in which all of the F-worlds are P-worlds. Since N does not permit E, all of the $E \vee F$-worlds in N must be F-worlds and, thus, P-worlds. In this case $(E \vee F) \blacksquare\!\!\rightarrow P$ is true at w as required. If (iii), let N′ be the E-permitting sphere in $\mathbb{N}_w$ in which all of the E-worlds are P-worlds. Since N′ does not permit F, all of the $E \vee F$-worlds in N must be E-worlds and, thus, P-worlds. In this case $(E \vee F) \blacksquare\!\!\rightarrow P$ is true at w as required. If (iv) let N be the F-permitting sphere in $\mathbb{N}_w$ in which all of the F-worlds are P-worlds and N′ be the E-permitting sphere in $\mathbb{N}_w$ in which all of the E-worlds are P-worlds. By nesting, either $N \subseteq N'$ or $N' \subseteq N$. In the former case, all of the E-worlds and all of the F-worlds in N must be P-worlds. In the latter case, all of the E-worlds and all of the F-worlds in N′ must be P-worlds. In either case $(E \vee F) \blacksquare\!\!\rightarrow P$ is true at w as required. **QED**

The validity of agglomeration reflects the fact that, if a piece of evidence normically supports each of two propositions, it must also normically support their conjunction. The validity of amalgamation reflects the fact that, if two pieces of evidence each normically support a proposition, it must also be normically supported by their disjunction.

In Chapter 6 I considered the absoluteness constraint for world normalcy according to which, for any two worlds w_1 and $w_2 \in W$, $\mathbb{N}_{w1} = \mathbb{N}_{w2}$. Absoluteness will serve to validate the following two inference patterns:

$$\frac{E \blacksquare\!\!\rightarrow P}{\Box(E \blacksquare\!\!\rightarrow P)} \qquad \frac{\sim(E \blacksquare\!\!\rightarrow P)}{\Box\sim(E \blacksquare\!\!\rightarrow P)}$$

Proof Suppose that $E \blacksquare\!\!\rightarrow P$ is true at a world $w \in W$. Either (i) there are no E-permitting spheres in $\mathbb{N}_w$ or (ii) there is an E-permitting sphere in $\mathbb{N}_w$ in which all of the E-worlds are P-worlds. If (i) then, by absoluteness, there are no E-permitting spheres in $\mathbb{N}_x$ for any $x \in W$. If (ii) then, by absoluteness there is an E-permitting sphere in $\mathbb{N}_x$ for every $x \in W$ in which all E-worlds are P-worlds. In either case $\Box(E \blacksquare\!\!\rightarrow P)$ is true at w. Suppose that $\sim(E \blacksquare\!\!\rightarrow P)$ is true at a world w. It follows that there is an E-permitting sphere in $\mathbb{N}_w$ and that all E-permitting spheres in $\mathbb{N}_w$ permit $E \wedge \sim P$. By absoluteness, for any $x \in W$, there is an E-permitting sphere in $\mathbb{N}_x$ and all E-permitting spheres in $\mathbb{N}_x$ permit $E \wedge \sim P$ in which case $\Box\sim(E \blacksquare\!\!\rightarrow P)$ is true at w. **QED**

With absoluteness in place, whether a normic support relation holds between two propositions will be a matter of necessity—normic support relations will be necessary relations. This is the view referred to in section 6.4 as *internalism* about normic support.

In fact, as the definition of the $\Box$ operator makes clear, a full absoluteness constraint is not required in order to deliver internalism about normic support. Both of the above inference patterns will be validated by what Lewis calls a *local absoluteness* constraint: for any worlds $w_1 \in W$ and $w_2 \in \cup ₦_{w1}$, $₦_{w1} = ₦_{w2}$. This constraint does not guarantee that every world imposes the same standards of normalcy, but it does guarantee that each world will deem only one standard of normalcy to be possible. While the full absoluteness constraint may better capture the thought that standards of normalcy are fixed or pre-ordained, logically speaking it adds nothing over and above the local absoluteness constraint (Lewis 1973: 122).

If we impose just the constraints suggested here, then the normic support relation will be governed by the logic **VTA** if we include absoluteness and **VT** if we do not (Lewis 1973: chap. 6).[6] There are, of course, many more possible constraints upon world similarity and world normalcy that we might investigate. It may be particularly interesting to consider constraints that govern the *relationship* between these two notions. Unlike the constraints that are intrinsic to world similarity or world normalcy, which validate inference patterns for either $\Box\!\!\rightarrow$ or $\blacksquare\!\!\rightarrow$, constraints of this kind will, in general, serve to validate 'mixed' inference patterns in which both operators feature. But I won't pursue this investigation any further here.

[6] Lewis does not provide axiomatisations for either of these logics in terms of the conditional. One possible axiomatisation of **VT** is as follows: we include, as axioms, all truth functional tautologies, definitions of $\Box$ and $\rightarrow$ in terms of $\blacksquare\!\!\rightarrow$ and all instances of the following schemas:

(1) $P \blacksquare\!\!\rightarrow P$
(2) $(\sim P \blacksquare\!\!\rightarrow P) \supset (Q \blacksquare\!\!\rightarrow P)$
(3) $(P \blacksquare\!\!\rightarrow \sim Q) \vee (((P \wedge Q) \blacksquare\!\!\rightarrow R) \equiv (P \blacksquare\!\!\rightarrow (Q \supset R)))$
(4) $\Box P \supset P$

The inference rules are modus ponens, the substitution of logical equivalents and Lewis's *Deduction With Conditionals* (multiple premise closure):

$$\text{For any } n \geq 1 \quad \frac{\vdash (P_1 \wedge \ldots \wedge P_n) \supset Q}{\vdash ((E \blacksquare\!\!\rightarrow P_1) \wedge \ldots \wedge (E \blacksquare\!\!\rightarrow P_n)) \supset (E \blacksquare\!\!\rightarrow Q)}$$

For **VTA**, we add all instances of the following schemas as axioms:

(5) $(P \blacksquare\!\!\rightarrow Q) \supset \Box (P \blacksquare\!\!\rightarrow Q)$
(6) $\sim(P \blacksquare\!\!\rightarrow Q) \supset \Box \sim(P \blacksquare\!\!\rightarrow Q)$

7.3. The Formal Structure of Justification

I have catalogued a number of notable formal features of normic support. According to what I've called the *normic theory* of justification, one has justification for believing a proposition P just in case one's evidence provides normic support for P. Given the normic theory, justification will inherit all and only the formal features of normic support—it will, in particular, exhibit closure under multiple premise deductive consequence, amalgamation, cumulative transitivity, and rational and cautious monotonicity. The question of whether the principle of multiple premise closure holds for justification is one that has been much discussed. Aside from this, though, relatively little attention has been paid to questions about the formal principles that justification ought to satisfy. In fact, a certain case can be made in favour of each of the principles just listed.

According to the principle of amalgamation for justification, if E provides justification for believing P and F provides justification for believing P then $E \vee F$ provides justification for believing P. Suppose one is about to conduct a test which must yield one of two possible results—either it will yield a body of evidence E or it will yield a body of evidence F. Suppose that, if one possesses evidence E it would be epistemically permissible for one to believe P and, if one possesses evidence F it would be epistemically permissible for one to believe P. Surely then, if one has full information about the test and the results it can possibly yield—that is, if one has the evidence $E \vee F$—then it should *already* be epistemically permissible for one to believe P. Surely, if one were interested only in taking a view about P, there is no need to go ahead with the test. But if amalgamation fails, then there is no guarantee that this will be so.

According to the principle of cumulative transitivity for justification, if E provides justification for believing P and $E \wedge P$ provides justification for believing Q then E provides justification for believing Q. This principle guarantees that whenever it is epistemically permissible, given a certain body of evidence, to believe a proposition P, it is also epistemically permissible to *add* that proposition to one's body of evidence in the service of drawing further conclusions. If cumulative transitivity holds then, by doing this, one will never be led to form beliefs that are not epistemically permitted, given one's initial evidence. This, in effect, is what legitimises reasoning in inferential chains—in using one's conclusions as the basis for further inferences. If cumulative transitivity fails, then there is no guarantee that such a procedure is legitimate—just because it is epistemically permissible to believe a proposition, it may yet be impermissible to use it as a basis for inference.

According to the principle of cautious monotonicity for justification if E provides justification for believing P and E provides justification for believing Q then $E \wedge Q$ provides justification for believing P. According to the principle of rational monotonicity for justification if E provides justification for believing P and E does not provide justification for believing ~Q then $E \wedge Q$ provides justification for believing P. Both the cautious and rational monotonicity principles seem to be required in order to enforce a coherent attitude towards *defeaters*. The cautious monotonicity principle could be phrased in this way: if E provides justification for believing P and $E \wedge D$ does not provide justification for believing P—that is, if D defeats E's justification for P—then E does not also provide justification for believing D. And the rational monotonicity principle could be phrased in this way: if E provides justification for believing P and $E \wedge D$ does not provide justification for believing P—if D defeats E's justification for P—then E must provide justification for believing ~D.

If the former principle fails then one's evidence could provide one with epistemic permission to believe a proposition P, vulnerable to defeat by D, and yet *also* provide one with epistemic permission to believe that D holds. If the latter principle fails then one's evidence could provide one with epistemic permission to believe a proposition P, vulnerable to defeat by D, and yet *also* make it impermissible to believe ~D. If one believes a proposition, while acknowledging that the permission to believe it is vulnerable to a certain defeater, then one should, and surely *would*, believe that the defeater does not hold. But if these monotonicity principles fail, then there is no guarantee that one would be epistemically permitted to believe this.

Much like the principle of multiple premise closure, the principles just considered appear to reflect ingrained aspects of our general reasoning practices. As with multiple premise closure, simply making this observation need not be the end of the story—there may be other considerations bearing upon their acceptability. But each of these principles does, at the very least, have a certain prima facie appeal.

Say that E *probabilistically supports* a proposition P, just in case the probability of P conditional upon E exceeds a threshold t—$Pr(P \mid E) > t$, for some t close to, but less than, 1. According to the risk minimisation conception of justification, one has justification for believing P just in case one's evidence probabilistically supports P in the way just defined. Given the risk minimisation conception, justification will inherit all and only the formal features of probabilistic support. It is well known that justification, on the risk minimisation conception, will not satisfy the principle of multiple premise closure. Given the result proved in section 4.2 it follows that justification, on the risk minimisation conception, must fail to satisfy either cumulative transitivity or cautious monotonicity. In fact, we can prove a lot more than this; on the risk minimisation conception, justification will satisfy *none* of the principles just considered (and, as such, the failure of multiple premise closure, on the risk minimisation conception, is effectively *overdetermined*).

We continue to model propositions as subsets of a set of possible worlds W, with conjunction being modelled as set theoretic intersection, disjunction as set theoretic union, and negation as complementation in W. A *probability space* consists of a set of worlds W, a set of propositions F that includes W itself and is closed under negation and disjunction, and a function Pr taking each proposition in F to a real number in the unit interval in a way that conforms to Kolmogorov's probability axioms:

(i) $Pr(P) \geq 0$
(ii) $Pr(W) = 1$
(iii) $Pr(P \vee Q) = Pr(P) + Pr(Q)$, for any P and Q such that P and Q are inconsistent.

Conditional probabilities are defined using the standard *ratio formula*— $Pr(P \mid Q) = Pr(P \wedge Q)/Pr(Q)$ if $Pr(Q) > 0$ and is undefined otherwise. If we are considering a 'maximal' probability space, W might be thought of as the set of all worlds that are metaphysically or epistemically possible. More often, in practice, W is thought of as a set of exclusive and exhaustive possibilities that may not be maximally specific—like the possible outcomes of a dice roll or lottery. Pr is to be understood, informally, as an evidential probability function—with Pr(P) representing the probability of P given a background body of evidence and $Pr(P \mid E)$ representing the evidential probability of P given E, along with background evidence.

A probability space is said to be *finite* if W has a finite number of members and *uniform* if Pr assigns the same probability to each of its members—for any $w \in W$, $Pr(\{w\}) = 1/|W|$ (where $|W|$ is the cardinality of W). In a finite, uniform probability space it follows from the axioms that the probability of any proposition will simply be equal to the ratio of its cardinality to that of W—for any $P \in F$, $Pr(P) = |P|/|W|$. Furthermore, in a finite, uniform probability space it follows from the ratio formula that the probability of one proposition P conditional upon another proposition Q will, if defined, be equal to the ratio of the cardinality of $P \wedge Q$ to that of Q—for any $P, Q \in F$, $Pr(P \mid Q) = |P \wedge Q|/|Q|$. To show that probabilistic support does not satisfy amalgamation, cumulative transitivity, or rational or cautious monotonicity, it suffices to consider finite, uniform spaces.

For ease, let the threshold t be set at 0.75. In this case E probabilistically supports P iff $Pr(P \mid E) > 0.75$. Consider a finite probability space in which W has 6 members $w_1, w_2 \ldots w_6$ corresponding to, say, the ways that a regular die might land. We let F consist of all the subsets of W and suppose that Pr gives a uniform distribution over the members of W—that is, $Pr(\{w_1\}) = Pr(\{w_2\}) = Pr(\{w_3\}) \ldots = 1/6$. Let $E = \{w_1 \ldots, w_5\}$, $F = \{w_2 \ldots, w_6\}$ and $P = \{w_2 \ldots, w_5\}$. $Pr(P \mid E) = Pr(P \mid F) = 4/5 > t$. E probabilistically supports P and F probabilistically supports P. $E \vee F = W = \{w_1 \ldots, w_6\}$ and $P \wedge (E \vee F) = P = \{w_2 \ldots, w_5\}$. As such $Pr(P \mid E \vee F) = 4/6 \leq t$. $E \vee F$ does not probabilistically support P. Probabilistic support does not satisfy amalgamation.

Now let $E = W = \{w_1 \ldots, w_6\}$, $P = \{w_1 \ldots, w_5\}$ and $Q = \{w_1 \ldots, w_4\}$. Notice that $E \wedge P = P = \{w_1 \ldots, w_5\}$, $E \wedge Q = Q = \{w_1 \ldots, w_4\}$ and $E \wedge P \wedge Q = Q = \{w_1 \ldots, w_4\}$. $Pr(P \mid E) = 5/6 > t$ and $Pr(Q \mid E \wedge P) = 4/5 > t$. E probabilistically supports P and $E \wedge P$ probabilistically supports Q. $Pr(Q \mid E) = 4/6 \leq t$. E does not probabilistically support Q. Probabilistic support does not satisfy cumulative transitivity.

Finally, let $E = \{w_1 \ldots, w_5\}$, $P = \{w_1 \ldots, w_4\}$, and $Q = \{w_2 \ldots, w_5\}$. Notice that $E \wedge Q = Q = \{w_2 \ldots, w_5\}$, $E \wedge {\sim}Q = \{w_1\}$, and $E \wedge P \wedge Q = P \wedge Q = \{w_2 \ldots, w_4\}$. $Pr(P \mid E) = Pr(Q \mid E) = 4/5 > t$. $Pr({\sim}Q \mid E) = 1/5 \leq t$. E probabilistically supports P and E probabilistically supports Q and E does not probabilistically support ~Q. $Pr(P \mid E \wedge Q) = 3/4 \leq t$. $E \wedge Q$ does not probabilistically support P. Probabilistic support does not satisfy cautious or rational monotonicity.

For these calculations, the value of t was set at 0.75—but the particular value we choose is of no consequence. Provided the value of t is greater than or equal to 0.5, and strictly less than 1, corresponding calculations will go through. For any t, $0.5 \leq t < 1$, let n be the smallest integer strictly greater than $1/(1 - t)$. To find counterexamples to amalgamation, cumulative transitivity, and cautious and rational monotonicity, we need only consider a finite, uniform probability space in which W has n + 1 elements, and then run the calculations as above.

It is very common to assess theories of justification by exploring their predictions about particular hypothetical cases. With the exception of multiple premise closure, it is much less common to assess theories of justification by exploring their predictions about formal principles. One significant benefit of treating a body of evidence as a trove of propositions, and of treating justification as a function of evidence, is that justification attributions can be cast in a *conditional* form. This greatly expands the range of formal principles for justification that we can consider, and enables us to capture many more aspects of our general reasoning practices with such principles. Only the multiple premise closure principle allows for the conditional form of justification attributions to be suppressed—and this may account for its relative prominence. From the perspective developed here, however, this feature is purely incidental, owing to nothing more than the fact that the antecedent evidence term happens to be constant throughout the principle.

The risk minimisation conception of justification forces us to sacrifice the principle of multiple premise closure. This fact is often observed, but may only be the tip of an iceberg. For our intuitions about the formal structure of justification run much deeper than our feelings about this one principle. And, on close scrutiny, the risk minimisation conception leaves very little, if any, of this intuitive formal structure intact.

Appendix to Chapter 7: Catalogue of Inference Patterns for the Normic Conditional

Invalid Patterns

$$\begin{array}{l} E \\ \underline{E \blacksquare\!\!\rightarrow P} \\ P \end{array}$$

Modus Ponens

$$\begin{array}{l} E \blacksquare\!\!\rightarrow P \\ \underline{P \blacksquare\!\!\rightarrow Q} \\ E \blacksquare\!\!\rightarrow Q \end{array}$$

Transitivity

$$\begin{array}{l} \underline{E \blacksquare\!\!\rightarrow P \qquad\quad} \\ (E \wedge F) \blacksquare\!\!\rightarrow P \end{array}$$

Strengthening the Antecedent

Valid Patterns

$$\begin{array}{l} E \blacksquare\!\!\rightarrow P \\ (E \wedge P) \blacksquare\!\!\rightarrow Q \\ \hline E \blacksquare\!\!\rightarrow Q \end{array}$$

Cumulative Transitivity

$$\begin{array}{l} E \blacksquare\!\!\rightarrow P \\ E \blacksquare\!\!\rightarrow Q \\ \hline (E \wedge Q) \blacksquare\!\!\rightarrow P \end{array}$$

Cautious Monotonicity

$$\begin{array}{l} E \blacksquare\!\!\rightarrow P \\ \sim(E \blacksquare\!\!\rightarrow \sim Q) \\ \hline (E \wedge Q) \blacksquare\!\!\rightarrow P \end{array}$$

Rational Monotonicity

$$\begin{array}{l} E \blacksquare\!\!\rightarrow P \\ E \blacksquare\!\!\rightarrow Q \\ \hline E \blacksquare\!\!\rightarrow (P \wedge Q) \end{array}$$

Agglomeration

$$\begin{array}{l} E \blacksquare\!\!\rightarrow P \\ F \blacksquare\!\!\rightarrow P \\ \hline (E \vee F) \blacksquare\!\!\rightarrow P \end{array}$$

Amalgamation

$$\begin{array}{l} E \blacksquare\!\!\rightarrow P \\ P \rightarrow Q \\ \hline E \blacksquare\!\!\rightarrow Q \end{array}$$

Weakening the Consequent

Optional Patterns

$$\begin{array}{l} E \blacksquare\!\!\rightarrow P \\ \hline \Box(E \blacksquare\!\!\rightarrow P) \end{array}$$

Necessitation1

$$\begin{array}{l} \sim(E \blacksquare\!\!\rightarrow P) \\ \hline \Box\sim(E \blacksquare\!\!\rightarrow P) \end{array}$$

Necessitation2

8

Introducing Degrees

In Chapter 7 I outlined a model for exploring the formal features of safety and normic support. If we accept what Lewis called the *limit assumption*, this will introduce sufficient structure into the model to define *degrees* of safety and of normic support. While my concern, thus far, has been with qualitative or outright belief, degrees of safety and normic support may provide a foundation for a certain epistemology of *degrees of belief*. The picture that emerges turns out to be at odds with the more familiar picture presented by Bayesian epistemology.

8.1. The Limit Assumption

In the formal model that I have described, propositions are treated as subsets of a set of possible worlds W. All of the epistemic information in the model is encoded by two functions \$ and ₦ taking possible worlds to systems of spheres. The sphere systems assigned by \$ are to be understood as carrying information about the comparative similarity of worlds while the sphere systems assigned by ₦ are to be understood as carrying information about the comparative normalcy of worlds. In addition to the compulsory constraints of non-emptiness, nesting, and closure under unions and non-empty intersections, I suggested that \$ be subject to a weak centering constraint and ₦ be subject (perhaps) to an absoluteness constraint. Another optional constraint upon sphere systems that Lewis considers is the so-called *limit assumption* according to which, for any world w and possible proposition P, there is a unique smallest P-permitting sphere associated with w. With propositions modelled as sets of possible worlds, what this amounts to is that, for any set of possible worlds that overlaps the largest sphere associated with w, there will be a unique smallest sphere associated with w that it overlaps.

If there are only finitely many spheres associated with w, then nesting guarantees that the limit assumption is satisfied, but if the number of spheres associated with w is infinite then this is not guaranteed. Both the centering and weak centering constraints yield one special case of the limit assumption—namely the existence of an *overall* smallest non-empty sphere associated with any world (this guarantees any instance of the limit assumption for which P is W). Lewis, in any case, is inclined to reject the limit assumption in full generality—at least when sphere systems are understood as carrying information about comparative world similarity. For Lewis, some propositions have the feature that there are no maximally similar worlds in which they are true.

Suppose I'm holding a ruler that is just under 35 cm long. Consider the proposition that the ruler is longer than 35 cm. According to Lewis, there are no maximally similar worlds in which this proposition holds—for any world in which this proposition is true we can imagine a *more similar* world in which it's true. If we consider a world in which the ruler is 35.1 cm long, it seems clear that an otherwise identical world in which it is 35.05 cm long would count as more similar and an otherwise identical world in which it is 35.025 cm long would count as more similar again and so on. According to Lewis, for any n, the most similar worlds in which the ruler is 35 + n cm long will be less similar than the most similar worlds in which the ruler is 35 + n/2 cm long.

By arguing in this way, though, Lewis appears to lose sight of some of his own advice on how to think about the notion of world similarity. As I noted in section 6.2, there are any number of different respects in which possible worlds might be said to be similar to one another and it is only by appropriately weighting these respects that we can devise an overall similarity measure of the sort that Lewis requires. But, rather than trying to make brute judgements about which respects of similarity and difference ought to be important, Lewis suggests that we let counterfactual conditionals themselves serve as our guide.

What is, I think, genuinely indisputable is that a world in which the ruler is 35.05 cm is more similar to the actual world, *in some respect*, than a world in which it is 35.1 cm and, in general, that a world in which the ruler is 35 + n/2 cm long is more similar to actuality in some respect than a world in which the ruler is 35 + n cm long for any n. What can be disputed, though, is whether this respect is to be assigned *any* weight when it comes to determining the overall similarity ordering that is relevant for the evaluation of counterfactuals. Consideration of the relevant counterfactuals does not appear to support a positive answer.

If the most similar worlds in which the ruler is 35 + n/2 cm long are more similar than the most similar worlds in which the ruler is 35 + n cm long, for any n, then, given Lewis's analysis, the following counterfactuals must all be true: 'If the ruler were longer than 35 cm, it would not be longer than 35.1 cm,' 'If the ruler were longer than 35 cm it would not be longer than 35.05 cm,' 'If the ruler were longer than 35 cm it would not be longer than 35.025 cm,' and so on . . . Such counterfactuals do not seem clearly true—if anything they become increasingly dubious as the value of n shrinks. Indeed, the entire infinite set of such counterfactuals is, in effect, *ω-inconsistent* (see Spohn 2012: 73–4) None of this is to say that the limit assumption should be accepted, but Lewis gives no clear reason for rejecting it. I will have a little more to say about it in section 8.2.

The limit assumption, as Lewis shows, is logically inert. That is, imposing the limit assumption will leave the logic of □→ intact—and the structure of the safe evidence relation unchanged.[1] But this is not to say that the limit assumption is a matter of *indifference*—on the contrary, it may turn out to have a profound significance for our concept of safety. Without the limit assumption, the system of similarity spheres associated with a world w might include infinite descending chains of smaller and smaller spheres that continue on without end. If we impose the limit assumption, then the spheres associated with w can always be well-ordered by size, and assigned a canonical ordinal numbering. The smallest sphere to overlap W will be S_1, the smallest sphere to overlap (W – S_1) will be S_2, the smallest sphere to overlap (W – S_2) will be S_3, and so on. If there are only finitely many spheres in the system, then we will reach the largest sphere $\cup\$_w$ within a finite number of steps. If there are infinitely many spheres, then there may be a sphere S_n for every positive integer n. Things may be taken even further than this. The union of all of the S_ns may still be smaller than $\cup\$_w$. This sphere could then be assigned the first transfinite ordinal ω. We then begin again; the smallest sphere to overlap (W – S_ω) will be $S_{\omega+1}$ and so on. We may find

[1] Strictly speaking, the limit assumption is logically inert *relative* to the class of constraints that Lewis considers (see Lewis 1973: chap. 6). That is, the logic that results from adding the limit assumption to any combination of constraints selected from this class is identical to the logic that results from the constraints alone. It is quite possible to construct constraints relative to which the limit assumption is not logically inert—but no such constraints will be considered here.

spheres corresponding to further limit ordinals—$\omega.2$, $\omega.3\ldots\omega^2$, $\omega^3\ldots$—and uncountable ordinals as well, though the cardinality of the set of possible worlds W (whatever we take it to be) will impose some restrictions on this.

If similarity spheres can be numbered then, whenever E stands in the safe evidence relation to P, it will be possible to attach a number to this relation, which might be thought to gauge its strength—namely, the number of the largest sphere throughout which all the E-worlds are P-worlds. If we impose the limit assumption, then we can begin to think about safety as something that comes in *degrees*. If we have similarity spheres corresponding to transfinite ordinals, then it will be possible also for evidence to confer transfinite ordinal degrees of safety. Though this possibility is worth noting, I will bracket it here by restricting myself to models in which sphere systems have at most one infinite ascending chain of spheres, with the largest sphere in the system as its least upper bound.

8.2. Degrees of Safety

Let E be a body of evidence possessed at world w. The degree of safety conferred upon proposition P by E at w can be defined as the number of the largest sphere in $\$_w$ throughout which all E-worlds are P-worlds. The degree of safety that E confers upon P at w will be one lower than the number of the smallest sphere in $\$_w$ that permits $E \wedge {\sim}P$, if there is such a sphere. The safety of a belief could also be assigned a degree corresponding to the degree of safety conferred upon its content by the evidence upon which the belief is based.

To tie up a few loose ends, the empty sphere, which has to be associated with every world, might be designated S_0. If the smallest non-empty sphere in $\$_w$ already includes worlds at which E is true and P is false, then the empty sphere will be the largest sphere in $\$_w$ in which all of the E-worlds are P-worlds and the degree of safety conferred upon P by E at w will be 0. At the other extreme, let the largest sphere in $\$_w$, $\cup\$_w$, be designated S_∞.[2] If all of the E-worlds in this sphere are P-worlds

[2] S_∞ is added here as a kind of *maximum* sphere size, to complement S_0's minimum. Just because S_∞ is the largest similarity sphere in $\$_w$, this is not to suggest, of course, that there need be an infinite number of spheres in $\$_w$. If there were only n spheres in $\$_w$, for finite n, then we could 'count our way up' to S_∞. By marking this sphere with '∞' we preserve the information that it is maximal at w—not a subset of any other sphere in $\$_w$.

(that is, if E $\rightarrow$ P is true at w) we might say that E confers an 'infinite' degree of safety upon P at w—the highest degree of safety that is possible. If we define degrees of safety in the way suggested, it will turn out that the degrees of safety conferred upon a set of propositions by a body of evidence at a given world will be described by a *completely minimative positive ranking function* (see Spohn 2012: 75).

A *positive ranking space* consists of a set of worlds W, a set of propositions F that includes W itself and is closed under negation and conjunction, and a function r taking each proposition in F to a member of $N \cup \{\infty\}$—that is, the set of non-negative integers plus infinity—in a way that conforms to the ranking axioms:

(i) $r(W) = \infty$
(ii) $r(\emptyset) = 0$
(iii) $r(P \wedge Q) = \min(r(P), r(Q))$ for any P, Q

If F is, in addition, infinite and closed under infinite conjunction then axiom (iii) can be strengthened to:

(iv) For any set of propositions Σ, $r(\wedge\Sigma) = \min\{r(P) \mid P \in \Sigma\}$

A positive ranking function that satisfies (iv) is termed *completely minimative.*

We prove in two stages that the degrees of safety conferred upon propositions by a body of evidence at a possible world conform to the axioms for a completely minimative positive ranking function. First, let the *degree of necessity* of a proposition P at a world w be equal to the number of the largest sphere in $\$_w$ throughout which P holds. Let the degree of necessity of P at w be equal to ∞ in case P holds throughout $\cup\$_w$ and equal to 0 in case P does not hold throughout any non-empty sphere in $\$_w$.

Degrees of necessity, so defined, will conform to the axioms for a completely minimative positive ranking function.

Proof The degree of necessity of W and $\emptyset$ at a world $w \in W$ will evidently be ∞ and 0 respectively. Degrees of necessity conform to axioms (i) and (ii). Suppose that a proposition P has a degree of necessity of n at w and a proposition Q has a degree of necessity of m at w, such that $m \geq n$. By the definition of degrees of necessity, S_m is the largest sphere in $\$_w$ throughout which Q holds, and S_n is the largest sphere in $\$_w$ throughout which P holds. Since $m \geq n$, it follows that $S_m \supseteq S_n$ in which case S_n is the largest sphere in $\$_w$ throughout which $P \wedge Q$ holds. In this case, by the definition of degrees of necessity, P ∧ Q has a degree of necessity of n at w. Degrees of necessity conform to axiom (iii).

More generally, suppose we have a set of propositions Σ. Let R(Σ) be the set of degrees of necessity of members of Σ at w. Since $R(\Sigma) \subseteq N \cup \{\infty\}$, this set can be well-ordered and must have a minimum member m. By the definition of degrees of necessity, the mth sphere in $\$_w$ is the largest sphere in $\$_w$ throughout which every proposition in Σ holds. In this case, the conjunction ∧Σ has a degree of necessity of m at w. Degrees of necessity conform to axiom (iv). **QED**

Let E be a body of evidence possessed at world w ∈ W. It follows, given weak centering, that the degree of necessity of ~E at w will be 0. The degree of safety that E confers upon a proposition P at w is, in effect, equal to the degree of necessity of E ⊃ P at w. For all propositions P, let r′(P) = r(E ⊃ P). All that remains to be proved is that r′ is a completely minimative positive ranking function whenever r is a completely minimative positive ranking function, such that r(~E) = 0.

Proof E ⊃ W is equal to (W – E) ∪ W which is, in turn, equal to W. As such, r(E ⊃ W) = r(W) = ∞ in which case r′(W) = ∞ as required. E ⊃ Ø is equal to (W – E) ∪ Ø which is equal to ~E. As such, r(E ⊃ Ø) = r(~E). By stipulation r(~E) = 0 in which case r′(Ø) = 0 as required. r′ conforms to axioms (i) and (ii).

r′(P ∧ Q) = r(E ⊃ (P ∧ Q)). Since E ⊃ (P ∧ Q) is equal to (E ⊃ P) ∧ (E ⊃ Q), r(E ⊃ (P ∧ Q)) = r((E ⊃ P) ∧ (E ⊃ Q)). Since r conforms to axiom (iii), r((E ⊃ P) ∧ (E ⊃ Q)) = min(r(E ⊃ P), r(E ⊃ Q)). As such r′(P ∧ Q) = min (r(E ⊃ P), r(E ⊃ Q)). By the definition of r′, r′(P ∧ Q) = min(r′(P), r′(Q)) as required. r′ conforms to axiom (iii). Since E ⊃ ∧Σ is equal to ∧{(E ⊃ P) | P ∈ Σ}, r(E ⊃ ∧Σ) = r(∧{(E ⊃ P) | P ∈ Σ}). Since r conforms to axiom (iv), r(∧{(E ⊃ P) | P ∈ Σ}) = min{r(E ⊃ P) | P ∈ Σ}. As such, r′(∧Σ) = min{r(E ⊃ P) | P ∈ Σ}. By the definition of r′, r′(∧Σ) = min{r′(P) | P ∈ Σ} as required. r′ conforms to axiom (iv). **QED**[3]

Propositions about the degree of necessity of propositions can themselves be located within the model. The proposition that P has a degree of

[3] The range of a ranking function can be extended to cover the entire ordinal hierarchy, with the axioms left as they are (and '∞' interpreted as a kind of 'absolute' infinity, greater than any ordinal). Such functions—close to what are sometimes called *conditional ordinal functions* (Spohn 1988, 2012: 72–3)—will describe the degrees of safety conferred upon propositions by a body of evidence, when degrees of safety are permitted to extend into the transfinite.

necessity of m will be equal to the set of worlds for which the largest similarity sphere throughout which P holds has a number of m. As such, propositions about degrees of necessity will themselves have degrees of necessity at possible worlds. It follows, of course, that propositions about the degree of safety that one proposition confers upon another will themselves have degrees of safety conferred upon them by propositions at possible worlds. One of the benefits of the present model is that the treatment of higher-order degrees of safety simply falls out of the treatment of first order degrees of safety in the same way that the treatment of $f(f(x))$ falls out of the treatment of $f(x)$ when f is a function with the same domain and range.

For all that's been said so far, degrees of safety carry only *ordinal* information—we cannot assume that the difference between safety degrees 1 and 2 is equal to the difference between safety degrees 5 and 6 or 100 and 101 etc. It is also important to stress that degrees of safety should be understood as world-relative and, as such, cannot be meaningfully compared across worlds. From the fact that E provides $safe_{17}$ evidence for P at w_1 and E provides $safe_3$ evidence for Q at w_2 we cannot conclude that E provides *safer* evidence for P at w_1 than for Q at w_2—for the two scales may be different. Intra-world safety comparisons will, however, be legitimate—from the fact that E provides $safe_{17}$ evidence for P and $safe_3$ evidence for Q at w we *can* conclude that E provides safer evidence for P than for Q at w.

Even if we reject the limit assumption, and deny that safety comes in degrees, intra-world safety comparisons will still be possible. We can say, in general, that E provides safer evidence for P than for Q at w just in case there is a sphere in $\$_w$ that includes worlds at which E is true and Q is false but does not include worlds at which E is true and P is false—a sphere that permits $E \wedge \sim Q$ but does not permit $E \wedge \sim P$. If we have the limit assumption in place, then this is equivalent to the condition that E confers a higher degree of safety upon P than Q at w—but such comparisons can be perfectly legitimate even if not supported by an underlying scale.

In section 6.3, I outlined a contextualist theory of knowledge attributions according to which a knowledge-of-P attribution is true at a world w only if the subject believes P and his belief is $safe_s$ where S is the smallest sphere in $\$_w$ to accommodate the truth of all salient propositions. This theory could be expressed in the language of degrees of safety. Let the *degree of far-fetchedness* of a proposition P at world w be equal to the number of the smallest similarity sphere in $\$_w$ to permit P (or equal to ∞ if there are no similarity spheres associated with w that permit P). If finite, the degree of far-fetchedness of a proposition will be one greater than the degree of necessity of its negation. If a sphere accommodates the truth of a salient proposition then it will *ipso facto* accommodate the truth of all propositions that are equally or less far-fetched. As such, a sphere will be the smallest to accommodate all salient propositions iff it is the smallest to accommodate the most far-fetched salient proposition.

With this is mind, the present contextualist theory could be expressed as follows: a knowledge-of-P attribution is true only if the subject believes P and the belief has a degree of safety that is greater than or equal to the degree of far-fetchedness of the most far-fetched salient proposition. Suppose a subject believes P based on E at w. Let S be the smallest sphere in $\$_w$ to accommodate the truth of all salient propositions. The subject's belief will be $safe_S$ iff all of the E-worlds in S are P-worlds. If m is the degree of far-fetchedness of the most far-fetched salient proposition, then S will have a number of m. By the definition of degrees of safety, all of the E-worlds in S will be P-worlds just in case the degree of safety that E confers upon P at w is greater than or equal to m.

8.3. Degrees of Normic Support

In section 2.4 I first mentioned an assumption to the effect that, for any possible proposition P and world w, there will be maximally normal worlds at which P holds from the perspective of w. This is, of course, nothing other than Lewis's limit assumption reconfigured for world normalcy rather than world similarity. As I noted, this assumption makes for a very simple definition of the normic support relation: evidence E normically supports proposition P iff P is true at all of the most normal worlds at which E is true. But the assumption has more far-reaching implications than this.

In the same way that the limit assumption for world similarity guarantees that the spheres assigned to a world by \$ can be well-ordered and numbered, the limit assumption for world normalcy guarantees that the spheres assigned to a world by ₦ can be well-ordered and numbered. The smallest sphere to overlap W will be N_1, the smallest sphere to overlap (W – N_1) will be N_2, the smallest sphere to overlap (W – N_2) will be N_3, and so on. Once again, this may be extended to transfinite ordinals—but I will set aside this possibility here.

Degrees of safety were defined as follows: the degree of safety that evidence E confers upon a proposition P at world w is equal to the number of the largest similarity sphere in $\$_w$ in which all of the E-worlds are P-worlds. The degree of safety that E confers upon P at w is one lower than the number of the smallest sphere in $\$_w$ that permits E ∧ ~P, if there is such a sphere. The degree of safety that E confers upon P is considered infinite in case all of the E-worlds in the largest sphere $\cup\$_w$ are P-worlds. This definition cannot, however, simply be carried over to normic support by replacing \$ with ₦—that is, the degree with which E normically supports P at w cannot simply be defined as the number of the largest sphere in $₦_w$ in which all of the E-worlds are P-worlds.

This definition proves problematic on account of the fact that there may be non-empty spheres in $₦_w$ that do not permit E—and these will count, trivially, as spheres in which all of the E-worlds are P-worlds. In general, the number of the largest sphere in $₦_w$ in which all of the E-worlds are P-worlds will be equal to the number of non-empty spheres that do not permit E plus the number of spheres that permit E and in which all of the E-worlds are P-worlds. But it's unclear why the former quantity should factor into the degree with which E normically supports P at w. The former quantity is in effect a measure of how abnormal *E* is at w—and, intuitively, this is irrelevant to the question of how strongly E normically supports P. On the present definition, the degree with which E normically supports P could in fact reach any degree n without E and P even *standing* in the normic support relationship. There could, after all, be n non-empty spheres in $₦_w$ that do not permit E and *no* E-permitting normalcy sphere in $₦_w$ in which all of the E-worlds are P-worlds.

In gauging the strength of the safety relation between E and P we are simply interested in how dissimilar the most similar E ∧ ~P worlds are—the less similar they are, the stronger the safety relation between E and P. In gauging the strength of the normic support relation between E and P, however, we're not interested in how abnormal *tout court* the most normal E ∧ ~P worlds are—this, as the above considerations illustrate, partly reflects the abnormality of E itself. What interests us is how much *more* abnormal the most normal E ∧ ~P worlds are than the most normal E-worlds—how many normalcy spheres separate the smallest sphere to permit E from the smallest sphere to permit E ∧ ~P.

Consider the following definition: the degree of normic support that evidence E confers upon proposition P at world w is equal to the number of the largest sphere in ₦$_w$ in which all E-worlds are P-worlds *minus* the number of the largest sphere in ₦$_w$ which does not permit E. The degree of normic support that E confers upon P at w will be equal to the number of the smallest sphere in ₦$_w$ to permit E ∧ ~P, if there is such a sphere, minus the number of the smallest sphere in ₦$_w$ to permit E. On this definition, E normically supports P iff the strength of the normic support relation between E and P has a positive degree.[4] The normic support of a belief could also be assigned a degree corresponding to the degree of normic support conferred upon its content by the evidence upon which it is based. As with similarity spheres, we can let N_∞ designate the largest normalcy sphere associated with a world w and say that the degree of normic support conferred upon P by E at w is infinite if all of the E-worlds in this sphere are P-worlds.[5]

Despite appearances, the definitions of degrees of safety and degrees of normic support do, in a sense, remain parallel. The definition of degrees of safety could be put in exactly the same form as the definition of degrees of normic support: the degree of safety that evidence E confers upon proposition P at world w is equal to the number of the largest sphere in $\$_w$ in which all of the E-worlds are P-worlds minus the number of the largest sphere in $\$_w$ which does not permit E. Weak centering, of course, guarantees that the largest sphere in $\$_w$ in which all the worlds

[4] This definition of degrees of normic support allows us to prove a satisfying result about the normic support of material conditionals. Consider three propositions E, P, and Q. Let a be the number of the smallest sphere in ₦$_w$ to permit E, b be the number of the smallest sphere in ₦$_w$ to permit E ∧ P, and c be the number of the smallest sphere in ₦$_w$ to permit E ∧ P ∧ ~Q. The degree with which E normically supports P ⊃ Q at w is equal to c − a. The degree with which E ∧ P normically supports Q at w is equal to c − b. Finally, the degree to which E normically supports ~P at w is equal to b − a. Since c − a = (c − b) + (b − a), the degree with which E normically supports P ⊃ Q at w is equal to the degree with which E normically supports ~P at w plus the degree with which E ∧ P normically supports Q at w. It's intuitive that there are two ways in which a body of evidence could normically support a material conditional—either by normically supporting the negation of the antecedent or by being *disposed* to normically support the consequent when supplemented by the antecedent. These are, in effect, the two components that sum to give the overall normic support for the conditional according to the above equation. I am indebted here to conversations with Brian Weatherson.

[5] If we allow the numbering of normalcy spheres to continue into the transfinite, then degrees of normic support may still be well defined—with 'minus' interpreted as the operation of *left subtraction*—and may take transfinite ordinal values. I won't explore this here. While the formalism does allow for the assignment of transfinite degrees of normic support, it is unclear whether such assignments can be sustained by the informal interpretation of normic support.

are ~E worlds is the empty sphere. This definition, then, is equivalent to the original definition given—weak centering, in effect, is what makes the simpler definition possible.

Is there any reason to accept the limit assumption for world normalcy? Lewis, as I noted in section 8.1, rejects the limit assumption in the case of world similarity. He argues against it by describing what he takes to be an infinite sequence of non-actual worlds, each one more similar to the actual world than the last. For reasons already discussed, I don't find the argument convincing—but even if one accepted Lewis's argument, there is no straightforward way to generalise it to the case of normalcy spheres. There is some temptation to suppose that world similarity permits of infinitely fine gradations—after all, there are respects of world similarity that clearly do have this feature. In contrast, there is no particular temptation to make the same supposition about world normalcy.

This does not, of course, amount to an *argument* in favour of the limit assumption for world normalcy. In a way, I have no arguments to offer in favour of this constraint—unless the profitable exploitation of an idea be thought to weigh in support of it, in which case this whole section might be seen as an elliptical argument. If we do see fit to impose the limit assumption, and come to think of normic support as coming in degrees, we open up a number of potentially fruitful avenues of inquiry that would otherwise remain closed to us.

Once normic support is thought of as something that comes in degrees, we should no longer think about *defeat* in a purely all-or-nothing way—rather, it too should be reconceived as something that permits of degree. The degree with which a proposition defeats a normic support relationship might be naturally defined as the diminution in normic support strength that it elicits. More formally, the degree with which D defeats the normic support relationship between E and P can be defined as the degree of the normic support relationship between E and P minus the degree of the normic support relationship between E $\wedge$ D and P (where $\infty - \infty = 0$). If N_a is the largest sphere of normalcy associated with w that does not permit E, N_b is the largest sphere of normalcy associated with w in which all E-worlds are P-worlds, N_c is the largest sphere of normalcy associated with w that does not permit E $\wedge$ D, and N_d is the largest sphere of normalcy associated with w in which all of the E $\wedge$ D-worlds are P-worlds, then the degree with which D defeats the normic support relation between E and P will be equal to $(b - a) - (d - c) = (b + c) - (d + a)$.

Given the definition of defeat offered in Chapter 7, D won't count as a defeater *simpliciter* for the normic support relation between E and P unless its degree is equal to that of the normic support relationship between E and P—that is, unless it serves to completely counteract the normic support that E affords P. If not, it will remain the case that E $\wedge$ D normically supports P—only more weakly than E alone. If E normically supports P then we might say that D is a *partial* defeater of this normic support relation iff the degree of normic support that E $\wedge$ D affords P is less than the degree of normic support that E affords P. We might say that D is a *full* defeater, or simply a defeater, of this normic support relation iff the degree of normic support that E $\wedge$ D affords P is 0.

It is possible to prove a kind of generalisation of the rational monotonicity principle for degrees of defeat: if E normically supports P, then the degree to which a proposition D defeats this normic support relationship can be no greater than the degree to which E supports ~D. That is, the degree to which E supports ~D places an upper limit on the extent to which D can defeat any of the normic support relations extending from E.

Proof Suppose E normically supports P at a world $w \in W$. Let N_a be the largest sphere of normalcy associated with w not to permit E, N_b be the largest sphere of normalcy associated with w in which all E-worlds are P-worlds, N_c be the largest sphere of normalcy associated with w not to permit E $\wedge$ D, and N_d be the largest sphere of normalcy associated with w in which all E $\wedge$ D-worlds are P-worlds. Since E normically supports P we have it that $b > a$. Since any sphere in which all E-worlds are P-worlds is also a sphere in which all E $\wedge$ D-worlds are P-worlds, we have it that $d \geq b$. The strength with which D defeats the normic support relation between E and P is, by the above definition, equal to $(b + c) - (d + a)$. The strength with which E normically supports ~D is equal to $c - a$. Since $d \geq b$ we have it that $(c - a) \geq (b + c) - (d + a)$ as required. **QED**

It follows from this that if E does not normically support ~D to any degree, then D cannot defeat any normic support relation extending from E to any degree, in line with the original rational monotonicity principle.

The definition of degrees of defeat, as it stands, allows for a proposition to defeat the normic support relationship with a *negative* degree. What this will indicate, in effect, is that the proposition serves to *reinforce* the normic support relationship in question—the negative degree of defeat indicating the strength with which it does so. We might say that the degree with which

R reinforces the normic support relationship between E and P can be defined as the degree of the normic support relationship between $E \wedge R$ and P minus the degree of the normic support relationship between E and P. Degrees of reinforcement can be thought of, then, as negative degrees of defeat and degrees of defeat as negative degrees of reinforcement. A 0 degree of defeat (or alternatively of reinforcement) could be taken to indicate a kind of *independence* between a proposition and a normic support relation.

Normic support, unlike safety, does not guarantee truth. From the fact that one possesses evidence E at a world w and that E normically supports P at w it does not follow that P is true at w. But the relationship between degrees of normic support and truth turns out to be somewhat more nuanced—it can be shown that there must be a threshold point such that truth *will* be guaranteed by any normic support relation of equal or greater degree—though this threshold point need not be constant across possible worlds.

Consider a world $w \in W$ and let m be the number of the smallest normalcy sphere associated with w that includes w. What m provides, in effect, is a measure of how abnormal w estimates itself to be—we might term this w's *self-abnormality index*. If w is a member of N_1 then m will be equal to 1, if it is a member of N_5, but no smaller spheres, m will be equal to 5, and so on. If w is not a member of any normalcy sphere associated with w, then the value of m might be set at infinity—but this circumstance is effectively ruled out by the total reflexivity constraint.

Suppose one possesses evidence E at world w. Let n be the number of the largest normalcy sphere associated with w which does not permit E. Notice that, since E is true at w, n must be less than m. In this case, $m - n$ serves as one's *truth threshold* for normic support at w. If the normic support conferred upon a proposition P by E at w is greater than or equal to $m - n$, then P must be true at w. The reasoning is straightforward: if the degree of normic support conferred on P by E at w is greater than or equal to $m - n$ then, since N_n is the largest normalcy sphere that does not permit E, the largest normalcy sphere in which all E-worlds are P-worlds must have a number greater than or equal to m. In this case, all of the E-worlds in N_m must be P-worlds. Since w is an E-world in N_m, w must be a P-world.[6] If

[6] If the smallest normalcy sphere to accommodate one's evidence is equal to the smallest normalcy sphere to include the world at which one is located, then one's truth threshold for normic support will diminish to 1. What this represents is a situation in which *every*

we were considering *similarity* spheres rather than normalcy spheres, then m would be equal to 1 for any $w \in W$—this is simply an alternative way of expressing the weak centering constraint, which also guarantees that n will be equal to 0 for any E true at w. As such, one's truth threshold for *safe* evidence is guaranteed to be 1 at any world—which is just to say that safety guarantees truth.

If we define degrees of normic support in the way suggested, then the degrees of normic support conferred upon a set of propositions by a body of evidence at a given world will, once again, conform to the axioms for a completely minimative positive ranking function:

(i) $r(W) = \infty$
(ii) $r(\emptyset) = 0$
(iii) $r(P \wedge Q) = \min(r(P), r(Q))$ for any P, Q
(iv) For any set of propositions Σ, $r(\wedge\Sigma) = \min\{r(P) \mid P \in \Sigma\}$

Let the *degree of entrenchment* of a proposition P at w be equal to the number of the largest normalcy sphere associated with w throughout which P holds. Let the degree of entrenchment of P at w be equal to ∞ in case P holds throughout the largest normalcy sphere associated with w and equal to 0 in case P does not hold throughout any non-empty normalcy sphere associated with w. Degrees of entrenchment, so defined, will conform to the axioms of a completely minimative positive ranking function. The proof is as the corresponding proof for degrees of necessity.

Let E be a body of evidence possessed at world w. It follows, given total reflexivity, that the degree of entrenchment of ~E at w will not be ∞. The degree of normic support that E confers upon a proposition P at w is equal to the degree of entrenchment of $E \supset P$ at w minus the degree of entrenchment of ~E at w (where $\infty - n = \infty$ for any $n \in N$). For all propositions P, let $r'(P) = r(E \supset P) - r({\sim}E)$. All that remains to be proved is that r' is a completely minimative positive ranking function whenever r is a completely minimative positive ranking function such that $r({\sim}E) \neq \infty$.

Proof Since $E \supset W$ is equal to W, $r(E \supset W) = r(W)$. In this case, $r'(W) = r(W) - r({\sim}E) = \infty$ as required. Since $E \supset \emptyset$ is equal to ~E, $r(E \supset \emptyset) = r({\sim}E)$.

abnormal feature of one's world is accounted for by one's evidence—a situation in which one's evidence is so rich that the world holds no more surprises in store. This can be thought of as a kind of 'end of inquiry' in which one's evidence can only normically support truths (though there may still be truths that it fails to normically support).

In this case $r'(\emptyset) = r(\sim E) - r(\sim E) = 0$ as required. r' conforms to axioms (i) and (ii). $r'(P \wedge Q) = r(E \supset (P \wedge Q)) - r(\sim E)$. Since $E \supset (P \wedge Q)$ is equal to $(E \supset P) \wedge (E \supset Q)$, $r(E \supset (P \wedge Q)) - r(\sim E) = r((E \supset P) \wedge (E \supset Q)) - r(\sim E)$. Since r conforms to axiom (iii) $r((E \supset P) \wedge (E \supset Q)) = \min(r(E \supset P), r(E \supset Q))$. As such $r'(P \wedge Q) = \min(r(E \supset P), r(E \supset Q)) - r(\sim E) = \min(r(E \supset P) - r(\sim E), r(E \supset Q)) - r(\sim E))$. By the definition of r', $r'(P \wedge Q) = \min(r'(P), r'(Q))$ as required. r' conforms to axiom (iii). Since $E \supset \wedge\Sigma$ is equal to $\wedge\{(E \supset P) \mid P \in \Sigma\}$, $r(E \supset \wedge\Sigma) - r(\sim E) = r(\wedge\{(E \supset P) \mid P \in \Sigma\}) - r(\sim E)$. Since r conforms to axiom (iv), $r(\wedge\{(E \supset P) \mid P \in \Sigma\}) = \min\{r(E \supset P) \mid P \in \Sigma\}$. As such, $r'(\wedge\Sigma) = \min\{r(E \supset P) \mid P \in \Sigma\} - r(\sim E) = \min\{r(E \supset P) - r(\sim E) \mid P \in \Sigma\}$. By the definition of r', $r'(\wedge\Sigma) = \min\{r'(P) \mid P \in \Sigma\}$ as required. r' conforms to axiom (iv). **QED**

One of the strengths of the present model is that propositions about the degree of entrenchment of propositions, like propositions about the degree of necessity of propositions, can be located within it. The proposition that P has a degree of entrenchment of m will be equal to the set $\{w \in W \mid$ the largest normalcy sphere associated with w throughout which P holds has a number of $m\}$. As such, propositions about degrees of entrenchment will themselves have degrees of entrenchment at possible worlds and it follows that propositions about the degree of normic support that one proposition confers upon another will have degrees of normic support conferred upon them by propositions at possible worlds.

It is important to point out that degrees of normic support, like degrees of safety, carry only ordinal information and may not be comparable across different possible worlds. From the fact that E provides a greater degree of normic support for P at w_1 than it provides for Q at w_2, it does not automatically follow that E normically supports P at w_1 more strongly than it normically supports Q at w_2—for the two scales may be different. There are, however, formal constraints—such as absoluteness and local absoluteness—which will make certain inter-world normic support comparisons legitimate. If absoluteness is imposed, then one can legitimately compare degrees of normic support across any two possible worlds. If local absoluteness is imposed, then one can legitimately compare degrees of normic support across any two worlds that share the same largest normalcy sphere. Intra-world normic support comparisons will, of course, always be legitimate.

Ranking theorists often make use of a notion of a *conditional rank*, intended to correspond to the notion of a conditional probability (see Spohn 2009: s. 2.1, 2012: s. 5.3). Positive conditional ranks are defined as follows: $r(P \mid Q) = r(Q \supset P) - r(\sim Q)$.[7] If r is a ranking function representing the degrees of entrenchment of propositions at a world w and r′ is a ranking function representing the degrees of normic support conferred upon propositions by a body of evidence E at w, we have it that $r'(P) = r(E \supset P) - r(\sim E) = r(P \mid E)$. The degree to which E normically supports P is equal to the conditional degree of entrenchment of P, given E.

According to the definition of reinforcing given above, R reinforces the normic support relationship between E and P just in case the strength with which $E \wedge R$ normically supports P is greater than the strength with which E alone normically supports P. If we let r be a ranking function representing the degree of entrenchment of propositions at a world $w \in W$, we have it that R reinforces the normic support relation between E and P just in case $r((E \wedge R) \supset P) - r(\sim(E \wedge R)) > r(E \supset P) - r(\sim E)$. By the definition of conditional ranks, $r(P \mid E \wedge R) = r((E \wedge R) \supset P) - r(\sim(E \wedge R))$. As such, we

[7] The relationship between conditional and unconditional ranks is, in a way, more perspicuous when presented in terms of *negative ranks*. A negative ranking function is a function nr taking propositions into $N \cup \{\infty\}$ that conforms to the following axioms:

(i) $nr(W) = 0$
(ii) $nr(\emptyset) = \infty$
(iii) $nr(P \vee Q) = \min(r(P), r(Q))$ for any propositions P, Q

A negative ranking function is deemed completely minimative if it also satisfies:

(iv) For any set of propositions Σ, $nr(\vee\Sigma) = \min\{nr(P) \mid P \in \Sigma\}$

A negative ranking function is one for which $nr(\emptyset)$ is set at ∞, nr(W) is set at 0, and the '$\wedge$'s in the third and fourth axioms for a positive ranking function are replaced with '$\vee$'s (see Spohn 2009: s. 2.1, 2012: chap. 5). As can be easily checked, negative ranks are simply the duals of positive ranks—if $nr(P) = r(\sim P)$ for all propositions P, then nr is a negative ranking function iff r is a positive ranking function. The normic bearing of a body of evidence could equally be represented by a negative ranking function, where the rank of a proposition is taken to signify the degree to which the evidence normically *refutes* it (normically supports its negation).

If nr is a negative ranking function we have it that $nr(P \mid Q) = nr(P \wedge Q) - nr(Q)$. This is a close counterpart of the standard ratio definition of conditional probabilities: if Pr is a probability function then $Pr(P \mid Q) = Pr(P \wedge Q)/Pr(Q)$. Let r be a positive ranking function and nr be the corresponding negative ranking function such that $r(P) = nr(\sim P)$ for all propositions P. By the definition of negative conditional ranks, $nr(\sim P \mid Q) = nr(\sim P \wedge Q) - nr(Q) = r(\sim(\sim P \wedge Q)) - r(\sim Q) = r(Q \supset P) - r(\sim Q)$. If we suppose that $r(P \mid Q) = nr(\sim P \mid Q)$ we arrive at the definition of positive conditional ranks given in the body text (see Spohn 2012: s. 5.3).

have the result that R reinforces the normic support relation between E and P just in case $r(P \mid E \wedge R) > r(P \mid E)$. Since partial defeat is defined as the opposite of reinforcing we also have it that D partially defeats the normic support relation between E and P just in case $r(P \mid E) > r(P \mid E \wedge D)$.

These characterisations of reinforcing and defeat effectively parallel the characterisations of confirmation and disconfirmation familiar from Bayesian confirmation theory. On the orthodox Bayesian picture, R counts as confirming P, relative to E just in case $Pr(P \mid E \wedge R) > Pr(P \mid E)$ and D counts as disconfirming P, relative to E just in case $Pr(P \mid E) > Pr(P \mid E \wedge D)$ where Pr is a prior probability function and conditional probability is defined according to the ratio formula: $Pr(P \mid Q) = Pr(P \wedge Q)/Pr(Q)$ (see, for instance, Earman 1992: chap. 3).

In section 6.3, I outlined a contextualist theory of justified belief attributions according to which a justified belief-in-P attribution is true at a world w only if the subject believes P and the belief is normically supported$_N$, where N is the smallest normalcy sphere associated with w to accommodate the truth of all salient propositions, and the evidence on which the belief is based is assumed to be one of these salient propositions. This theory could be expressed in the language of degrees of normic support. Let the *degree of abnormality* of a proposition P at world w be equal to the number of the smallest normalcy sphere associated with w to permit P (or equal to ∞ if there are no normalcy spheres associated with w to permit P). If finite, the degree of abnormality of a proposition will be one greater than the degree of entrenchment of its negation. If a normalcy sphere accommodates the truth of a salient proposition then it will *ipso facto* accommodate the truth of all propositions that are equally or more normal. As such, a normalcy sphere will be the smallest to accommodate all salient propositions iff it is the smallest to accommodate the most abnormal salient proposition.

With this in mind, the present contextualist theory could be expressed as follows: a justified belief-in-P attribution is true only if the subject believes P and the belief has a degree of normic support that is greater than the degree of abnormality of the most abnormal salient proposition minus the degree of abnormality of the evidence upon which the belief is based. Suppose a subject believes P based on E at w. Let N be the smallest normalcy sphere associated with w to accommodate the truth of all salient propositions. P will be normically supported$_N$ by E iff all of the E-worlds in N are P-worlds. If m is the degree of abnormality of the most abnormal salient proposition, then N will have a number of m. By the definition of degrees of normic support, all of the E-worlds in N will be P-worlds just in case the degree of normic support that E confers upon P at w plus the degree of abnormality of E at w is greater than m.

8.4. Degrees of Belief

Ranking functions are sometimes put forward as a way of representing coherent degrees of belief (see Spohn 2009, 2012, Huber 2009: s. 4)—an alternative to the more familiar Bayesian representation using probability functions. On the orthodox Bayesian picture, one's degrees of belief can be modelled by real numbers in the unit interval with 1 representing subjective certainty of truth, 0 representing subjective certainty of falsehood, and 0.5 representing perfect agnosticism. One's degrees of belief will count as coherent, for the Bayesian, insofar as they conform to the probability axioms:

(i) $Pr(P) \geq 0$
(ii) $Pr(W) = 1$
(iii) $Pr(P \vee Q) = Pr(P) + Pr(Q)$, for any P and Q such that P and Q are inconsistent.

If the domain of the function is taken to be infinite, and closed under countable disjunction, then (iii) may be strengthened:

(iv) If P_i, i = 1, 2 . . . is a sequence of pairwise inconsistent propositions then $Pr(\vee_i P_i) = Pr(P_1) + Pr(P_2) + Pr(P_3) \ldots$

Axiom (iv) is equivalent to axiom (iii) combined with the following *continuity* axiom:

(v) If P_i i = 1, 2 . . . is an increasing sequence of propositions ($P_1 \subseteq P_2 \subseteq P_3 \ldots$) then $Pr(\vee_i P_i) = \lim_{i \to \infty} Pr(P_i)$.

On the ranking theoretic picture, one's degrees of belief can be represented by natural numbers, and sometimes ∞, and will count as coherent insofar as they conform to the ranking axioms. For the ranking theorist, a degree of belief of ∞ represents subjective certainty of truth and a degree of belief of 0 represents the absence of any degree of belief. Importantly, if one invests a 0 degree of belief in a proposition P, this need not indicate subjective certainty of P's falsehood—indeed, it need not indicate any degree of *disbelief* in P. One's degree of disbelief in P, if we wish to speak in this way, will simply correspond to one's degree of belief in ~P. As can be easily checked, it is perfectly consistent with the ranking axioms that one assign a 0 degree of belief to both P and ~P—an assignment that might be thought to represent perfect agnosticism on the ranking theoretic representation.

The Bayesian framework is sometimes thought to have difficulty in accommodating our ordinary notion of qualitative or outright belief. Bayesians standardly take outright belief to correspond to an interval lying close to the top of the degree of belief scale. On this view there will be some threshold value t close to but less than 1, such that one counts as believing P iff one's degree of belief in P exceeds t (see, for instance, Foley 1992, 1993: chap. 4, 2009, Hawthorne and Bovens 1999, Sturgeon 2008, Hawthorne 2009). On this proposal, the categories of belief, disbelief, and agnosticism merely represent a coarse, threefold scheme for dividing up a more finely graded underlying scale. One consequence of this proposal, however, is that one's beliefs need not be closed under multiple premise deductive consequence—even when one's underlying degrees of belief are coherent. Suppose P and Q together entail R. One can coherently assign both P and Q a degree of belief greater than t and assign R a degree of belief less than t—such an assignment is consistent with the probability axioms. As such, one can coherently believe P, believe Q, accept that P and Q together entail R, and yet remain agnostic about R. Relatedly, one can coherently assign a degree of belief above t to each member of an inconsistent set of propositions—provided the set is sufficiently large, such an assignment need not contravene the probability axioms.[8] Coherent degrees of belief, on the Bayesian picture, need not give rise to consistent outright beliefs.

There are two slightly different ways in which outright belief might be introduced into the ranking theoretic framework: according to the first, one counts as believing a proposition just in case one invests some

[8] The minimum number of mutually inconsistent propositions which one can coherently believe will be determined, on the Bayesian picture, as a function of the belief threshold t. If one counts as believing P when one's degree of belief in P exceeds t, then one can coherently believe each member of an inconsistent set of propositions just in case it contains more than 1/(1 – t) members. That is, if I assign a degree of belief greater than t to each member of an inconsistent set of propositions, then this assignment can only be consistent with the probability axioms provided there are more than 1/(1 – t) propositions in the set (see Hawthorne and Bovens 1999: Theorem 1, Case 2).

positive degree of belief in it. According to the second, one counts as believing a proposition just in case one's degree of belief in the proposition exceeds some threshold $t > 0$ (see Spohn 2009: s. 2.1, 2012: s. 5.2). On the first proposal, one will only count as agnostic about P if one invests a 0 degree of belief in both P and its negation. On the second proposal the part of the degree of belief scale lying between 0 and t inclusive will represent a kind of agnosticism interval. One will count as agnostic about P if one's degree of belief assignments to P and to ~P fall within this interval. Being agnostic about P, on this proposal, is compatible with investing a positive degree of belief in either its truth or falsity.

On either of these proposals, one's outright beliefs are guaranteed to be closed under multiple premise deductive consequence, provided that one's underlying degrees of belief are coherent. Suppose P and Q together entail R. One cannot coherently assign both P and Q a degree of belief greater than t and assign R a degree of belief less than t—such an assignment is in conflict with the ranking axioms. Since P and Q entail R, $P \wedge Q \wedge R$ is equal to $P \wedge Q$ in which case $r(P \wedge Q \wedge R) = r(P \wedge Q)$. By axiom (iii) $r(P \wedge Q) = \min(r(P), r(Q))$ and $r(P \wedge Q \wedge R) = \min(r(P), r(Q), r(R))$. We have it that $\min(r(P), r(Q)) = \min(r(P), r(Q), r(R))$ in which case $r(R) \geq \min(r(P), r(Q))$ as required.

Furthermore, one cannot coherently assign a positive degree of belief to each member of an inconsistent set of propositions—any such assignment will contravene the ranking axioms. Suppose, for reductio, that we have a finite inconsistent set of propositions $\{P_1 \ldots P_n\}$ such that, for every i, $1 \leq i \leq n$, $r(P_i) > 0$. By $n - 1$ applications of axiom (iii) we have it that $r(P_1 \wedge \ldots \wedge P_n) > 0$. But, by stipulation, $P_1 \wedge \ldots \wedge P_n$ is equal to $\emptyset$ in which case $r(\emptyset) > 0$ contrary to axiom (ii). With axiom (iv) in place, a stronger result is available: for any set Σ of propositions such that $\wedge\Sigma = \emptyset$ there is some $P \in \Sigma$ such that $r(P) = 0$. On the ranking theoretic picture, coherent degrees of belief must give rise to a body of beliefs that is both consistent and deductively closed. This is often advertised as an advantage of the ranking theoretic framework over its Bayesian rival.

Any formal representation of degrees of belief must ultimately be made applicable to actual subjects. That is, the proponent of a particular formal representation of degrees of belief must ultimately supply some criteria by which to judge when a subject's degree of belief in a proposition is to be represented by one or other of the values permitted by the formalism. Bayesians standardly propose that numerical degrees of belief can be read off a subject's *betting dispositions*. Consider a bet that pays a certain reward in the event that P is true and nothing in the event that P is false. On the Bayesian proposal, at a first pass, one's degree of belief in P corresponds to the proportion of that reward that one would deem a fair price for the bet—that is, if one deems $1 a fair price for a bet that pays $2 in the event that P is true, then one's degree of belief in P is 0.5, if one deems $1.50 a fair price for this bet, then one's degree of belief in P is 0.75 etc. (see, for instance, Ramsey 1926, Jeffrey 1956). There are a number of well-known objections to this 'betting interpretation' of degrees of belief—some of which can be met by complicating or qualifying it in various ways, while others are more difficult to accommodate (see, for instance, Eriksson and Hájek 2007). One worry, which may go to the heart of the view, is that believing something seems to go hand in hand with a disposition to assert it and to use it as a premise in practical and theoretical reasoning. But one could lack any such disposition and still be willing to stake a large proportion of any potential reward for a bet on P.

When it comes to the ranking theoretic representation of degrees of belief, there is no standard story as to how these degrees might be manifested. One proposal begins from the very idea that proved problematic for the betting interpretation—that believing something involves a disposition to assert it and act upon it—and then identifies one's degree of belief with the tenacity or resilience of this disposition, in the face of contrary evidence. On this proposal, more precisely, one's degree of belief in P corresponds to the number of independent and minimally reliable sources telling against P that one would need to be exposed to in order to lose the disposition to assert and act upon P (Huber 2009: s. 4). Like the betting interpretation, this interpretation of degrees of belief would no doubt need to be complicated and qualified in various ways in order to be made viable. I won't attempt to defend this, or any, interpretation of degrees of belief here.

It is often said that one should proportion one's belief to the strength of one's evidence—that the degree to which one believes a proposition should match the degree to which one's evidence supports it (see Carnap 1950, Hempel 1965, Williamson 2000: chap. 8). This would seem to presuppose that degrees of belief and degrees of evidential support can be measured in the same way. If we do suppose that degrees of normic support and coherent degrees of belief can both be represented by natural numbers, and sometimes ∞, assigned in a way that conforms to the ranking axioms, then the following is one natural way to capture this requirement: one has justification for investing a degree of belief m in a proposition P iff one's evidence normically supports P to a degree of at least m.[9]

If one counts as believing P whenever one invests a positive degree of belief in P, then the normic theory of justification, as outlined in section 4.1, emerges as a limiting case of this proposal. On this view, one will have justification for believing P iff one's evidence normically supports P to a positive degree—which is just to say that one's evidence normically supports P. If one counts as believing P whenever one invests a degree of belief in P that exceeds a threshold $t > 0$, then (a version of) the threshold normic theory of justification, as outlined in section 5.2, emerges as a limiting case of the proposal. On this view, one will have justification for believing P iff P has a normic support degree that exceeds t, given one's evidence.

Looking at things in a slightly different way, adopting the ranking theoretic representation of degrees of belief provides a means of enriching or augmenting normic theories of justification—bringing degrees of belief, as well as outright belief, within their ambit. This could, perhaps, make for a powerful and cohesive overall view. But to fully explain and motivate such a view would require much. Amongst other things it would need to be squared with a particular interpretation of degrees of belief, such as that outlined above. This is more than I can attempt here.

[9] If we take a person's degree of belief in a proposition P to correspond to the number of independent, minimally reliable sources of information telling against P that the person could tolerate without losing their disposition to assert and act on P then, assuming one could only ever encounter a finite number of such sources, degrees of belief could only take natural numerical values, plus a single infinite value—∞—to represent a complete imperviousness to contrary evidence. If we then take degrees of normic support to provide a normative standard for degrees of belief, so understood, we have some motivation for limiting degrees of normic support to these values as well, as suggested in n. 5.

On the other hand, even if it isn't practically possible for a person to encounter an infinite number of minimally reliable sources of information telling against P within a finite time, such a thing may be metaphysically possible. If someone encountered, say, one such source in the first minute, a second in the next 30 seconds, a third in the next 15 seconds, a fourth in the next 7.5 seconds, and so on then, by the time two minutes had elapsed, one would have been exposed to an infinite number of such sources. One might still be willing to assert and act on P at the end of this process but, having gone through it, might now be prepared to give up this disposition if, say, just two more sources telling against P were encountered. This might provide grounds for reckoning the person's original degree of belief in P to have been $\omega + 2$. I won't pursue this line of thought further here. I am indebted here to conversations with Adam Rieger.

9

Refining Risk Minimisation

The Impossibility Results

One of the things that is puzzling about the widespread acceptance of the risk minimisation conception of justification is that the view is incompatible with a principle that has also long enjoyed widespread acceptance—namely, the principle of multiple premise closure. As mentioned in section 3.1, some philosophers have attempted to modify or refine the risk minimisation conception in such a way as to reconcile it with multiple premise closure. Many more, I suspect, have simply assumed that something like this could be done. The results proved in this final chapter raise serious doubts as to whether such a reconciliation is possible, even in principle. As such, those who are attracted to risk minimisation and to multiple premise closure cannot have things both ways and may be forced to decide where their ultimate loyalties lie.

9.1. The Low Risk Rule

According to the risk minimisation conception, high evidential probability suffices for justification: If $Pr(P) > t$, then one has justification for believing P, where Pr is one's evidential probability function and t is a real number close to, but less than, 1. Call this the *low risk rule*. The low risk rule is just one part of the risk minimisation conception as I've characterised it—but it is already enough to conflict with the principle of multiple premise closure: if one has justification for believing each of a set of propositions P_1, P_2, P_3 . . . , and P_1, P_2, P_3 . . . together entail Q, then one has justification for believing Q. The low risk rule and the principle of multiple premise closure are incompatible or, a little more carefully, they are together incompatible with the following principle: one cannot have justification for believing a contradictory proposition. This incompatibility can be made vivid via the lottery paradox.

Select an integer $n > 1/(1 - t)$ and consider a lottery such that it's part of my evidence that the lottery has n tickets, is fair, and exactly one ticket has won. The propositions that ticket #1 has lost, that ticket #2 has lost . . . that ticket #n has lost will each have a probability, given my evidence, of $1 - 1/n$ which, given the above inequality, will be greater than t. By the low risk rule, I have justification for believing each. By multiple premise closure, I have justification for believing that all the tickets have lost, which is inconsistent with the proposition that one ticket has won—a proposition that is certain, given my evidence. By a further application of multiple premise closure, then, I have justification for believing the out and out contradiction that all the tickets have lost and some ticket has won.

A lottery presents us in effect with a partition of possible outcomes, each of which is just as likely to obtain as the others. Believing, of a particular ticket, that it has lost the lottery is tantamount to dismissing one member of this partition. If the partition is large enough then the low risk rule will supply me with justification for dismissing each individual member. Multiple premise closure will then supply me with justification for simultaneously dismissing *all* the members of the partition—in spite of my knowing full well that one of these members must obtain.

While some risk minimisation theorists have responded to the lottery paradox by rejecting the principle of multiple premise closure, others have attempted to retain closure and to resolve the paradox by modifying or refining the low risk rule in such a way that it no longer supplies justification for believing lottery propositions. The refined rules that have been proposed can all be shoe-horned into the following form:

Refined Rule If $Pr(P) > t$, then one has justification for believing P *unless* P satisfies C.

where C is supposed to be a condition that is somehow distinctive of lottery propositions. The idea behind these rules is that high evidential probability *defeasibly* suffices for justification—suffices for justification provided the proposition in question is not exempted by condition C.

Consider the following rules, proposed by Pollock (1990: 80–1) and by Lehrer (1974: chap. 8):

Pollock's Rule If $Pr(P) > t$ then one has justification for believing P unless P is a member of a set of propositions Σ such that (i) Σ is *minimally inconsistent* and (ii) for any proposition Q that is a member of Σ, $Pr(Q) > t$.

Lehrer's Rule If $\Pr(P) > t$, then one has justification for believing P unless there is a *competitor proposition* Q such that $\Pr(Q) \geq \Pr(P)$.

A set of propositions is said to be minimally inconsistent just in case it is inconsistent, given one's evidence, and has no proper subset that has this feature. A proposition Q is said to be a competitor of a proposition P just in case the evidential probability of P conditional upon Q is lower than the unconditional evidential probability of P – $\Pr(P \mid Q) < \Pr(P)$.

The exemption conditions described by Pollock and Lehrer are clearly satisfied by lottery propositions. Each lottery proposition has an evidential probability of $1 - 1/n > t$. The set of lottery propositions is clearly minimally inconsistent. Furthermore, the lottery propositions are all competitors—the evidential probability of one lottery proposition conditional upon any other is equal to $1 - 1/(n - 1) < 1 - 1/n$. As such, every lottery proposition will have equally likely competitors. Pollock's exemption condition seems explicitly tailored to lottery propositions, while Lehrer's is less obviously so—and, as such, may appear somewhat less ad hoc. Both Pollock's and Lehrer's rules, in any case, avoid the lottery paradox—at least as it stands.

Consider now an arbitrary proposition P with an evidential probability of r such that $t < r < 1$. Select an integer n such that $n > 1/(1 - r)$ and consider a lottery such that it's part of my evidence that the lottery has n tickets, is fair, and exactly one ticket has won. Suppose further that the outcome of this lottery is probabilistically independent, given my evidence, of whether or not P is true. Let O_1 be the proposition that ticket #1 has won the lottery, O_2 be the proposition that ticket #2 has won the lottery, O_3 be the proposition that ticket #3 has won the lottery . . . O_n be the proposition that ticket #n has won the lottery. If Pr is my evidential probability function, then $\Pr(O_1) = \Pr(O_2) = \Pr(O_3) \ldots = \Pr(O_n) = 1/n$. Proposition P is equivalent to $(P \wedge O_1) \vee (P \wedge O_2) \vee (P \wedge O_3) \vee \ldots \vee (P \wedge O_n)$. Since the outcome of the lottery is probabilistically independent of P, $\Pr(P \wedge O_i) < 1/n$ and $\Pr({\sim}(P \wedge O_i)) > 1 - 1/n$, for any i, $1 \leq i \leq n$. Given the above inequality $\Pr({\sim}(P \wedge O_i)) > r > t$. The set $\{{\sim}(P \wedge O_1), {\sim}(P \wedge O_2), {\sim}(P \wedge O_3), \ldots, {\sim}(P \wedge O_n), P\}$ is minimally inconsistent and every member has an evidential probability greater than t. Thus, Pollock's exemption condition is satisfied by P.

Furthermore, ${\sim}(P \wedge O_1)$ is a competitor to P. By the ratio formula, $\Pr(P \mid {\sim}(P \wedge O_1)) = \Pr(P \wedge {\sim}(P \wedge O_1))/\Pr({\sim}(P \wedge O_1))$. Since $P \wedge {\sim}(P \wedge O_1)$ is equivalent to $P \wedge {\sim}O_1$, $Pr(P \wedge {\sim}(P \wedge O_1))/Pr({\sim}(P \wedge O_1)) = Pr(P \wedge {\sim}O_1)/Pr({\sim}(P \wedge O_1))$. Since P and O_1 are independent, $Pr(P \wedge {\sim}O_1)/Pr({\sim}(P \wedge O_1)) = Pr(P).Pr({\sim}O_1)/Pr({\sim}(P \wedge O_1))$. Since $Pr({\sim}O_1)/Pr({\sim}(P \wedge O_1)) < 1$, $Pr(P).Pr({\sim}O_1)/Pr({\sim}(P \wedge O_1)) < Pr(P)$ as required. Given that $Pr({\sim}(P \wedge O_1)) > Pr(P)$, Lehrer's exemption condition is satisfied for P. Neither Pollock's nor Lehrer's rule can supply justification for believing P. But P was nothing more than an *arbitrary* proposition with an evidential probability value stipulated to be somewhere between t and 1. Far from selectively exempting lottery propositions, Pollock's and Lehrer's rules would appear to exempt *any* proposition that is less than evidentially certain. Put slightly differently, Pollock's and Lehrer's rules turn out to be nothing more than roundabout reformulations of the following:

No Risk Rule If $Pr(P) = 1$, then one has justification for believing P.

It should come as no surprise that a rule like this avoids the lottery paradox—it does so at the cost of being trivial and uninformative.[1] This kind of argument traces back to Lehrer's own work (see Lehrer 1980).

The ambition behind refined rules like Pollock's and Lehrer's is not just to avoid the lottery paradox per se. Generally speaking, a rule expressing a sufficient condition for justification aspires to do two things: (i) supply justification for believing some propositions that are less than evidentially certain and (ii) avoid supplying justification for believing any contradictory propositions, even in combination with multiple premise closure. The low risk rule, of course, fails on the second count. Pollock's and Lehrer's rules, however, turn out to fail on the *first*—though this is scarcely obvious at first blush.

The failure of these two particular rules may be of little consequence—but the arguments used against them seem to gesture towards a much more general problem. What these arguments show is that *any* proposition that is less than certain can be completely divided into a series of subcases that are at least as unlikely as its negation. Believing the proposition, when viewed in this way, can seem like a particularly ill-fitting sort of commitment to take on. Believing the proposition then

[1] Strictly speaking, it is only relative to *finite* probability spaces that this rule should be regarded as trivial and uninformative. Relative to infinite probability spaces, the rule could reasonably be denied (see Chapter 2 n. 13). Nevertheless, it clearly supplies us with far less than Lehrer's and Pollock's rules are intended to.

seems tantamount to arbitrarily dismissing one possibility in a partition, even though the others are equally or less likely to obtain—much like dismissing one particular lottery outcome. The arguments against Pollock's and Lehrer's rules suggest that there is no deep probabilistic feature that sets lottery propositions apart from uncertain propositions more generally, and no prospect of finding a probabilistic condition that will selectively exempt them. One may get the impression that, so long as we restrict ourselves to the resources of probability, we will never find a refined rule that satisfies (i) and (ii).

This impression would seem to be borne out by an ingenious argument developed by Igor Douven and Timothy Williamson in 'Generalizing the lottery paradox' (2006). Douven and Williamson's argument purports to show that a strikingly broad range of refined rules—essentially all of those characterised in probabilistic terms—must either fail on count (i) or on count (ii). This is their 'generalized lottery paradox' and it comes close, I think, to showing that the ambition behind refined rules simply cannot be realised.[2]

9.2. The First Impossibility Result: Finite Probability Spaces

As in previous chapters, let propositions be modelled as subsets of a set of possible worlds W, with conjunction modelled as intersection, disjunction modelled as union, negation modelled as complementation in W, and entailment modelled as the subset relation. A probability space, then, is a triple $\langle W, \mathsf{F}, Pr\rangle$ where F is a set of propositions that includes W itself and is closed under negation and countable disjunction and Pr is an evidential probability function taking F into the set of real numbers in the unit interval. Douven and Williamson assume that $\langle W, \mathsf{F}, Pr\rangle$ is both *finite* and *uniform*. That is, they assume that W is a finite set and that Pr is a uniform distribution over its members—for any $w \in W$, $Pr(\{w\}) = 1/|W|$

[2] Further refined rules are presented by Douven (2002) and, arguably, by Ryan (1996). In more recent work, Lin and Kelly (2012), Arló-Costa and Pedersen (2012), and Leitgeb (2014b) identify probabilistic properties of propositions that are closed under multiple premise deductive consequence and entail probability above a set threshold. Attempts to put forward these properties as non-trivial sufficient probabilistic conditions for justification will also fall within the scope of the results to be proved. While the philosophers mentioned do have something like this goal in mind, their views are more circumspect. I shall return to this in n. 8.

(where $|W|$ is the cardinality of W). With these two assumptions in place, it follows that the probability of any proposition in F will be equal to the ratio of its cardinality to that of W—that is, for any $P \in F$, $Pr(P) = |P|/|W|$. With W a finite set, F can, for ease, be assumed to be the power set of W – $\wp(W)$.

Call a function f an *automorphism* of $\langle W, F, Pr \rangle$ iff f is a 1:1 function from F onto itself that satisfies three conditions:

(i) $f(P \vee Q) = f(P) \vee f(Q)$
(ii) $\sim f(P) = f(\sim P)$
(iii) $Pr(P) = Pr(f(P))$

for all $P, Q \in F$. An automorphism is a function mapping propositions to propositions in such a way as to preserve probabilities and logical relations.

Say that a property φ of propositions is *structural* with respect to a probability space $\langle W, F, Pr \rangle$ just in case, for any proposition $P \in F$ and automorphism f of $\langle W, F, Pr \rangle$, P has φ iff $f(P)$ has φ. A property φ of propositions is structural *simpliciter* just in case it is structural with respect to all probability spaces. A property φ of propositions is *aggregative* with respect to a probability space $\langle W, F, Pr \rangle$ just in case for any propositions $P, Q \in F$, $P \wedge Q$ has φ whenever P has φ and Q has φ. A property φ of propositions is aggregative *simpliciter* just in case it is aggregative with respect to all probability spaces. It's important to note that whether a proposition possesses a given property is also something that is probability space relative—a proposition may possess a property φ relative to some spaces in which it features, but not others. When it is obvious which probability space we are dealing with, this relativity can be suppressed (and Douven and Williamson do suppress it)—but it will assume some significance in section 9.3. Given these definitions, Douven and Williamson prove the following:

Impossibility 1 Let $<W, \wp(W), Pr>$ be a finite, uniform probability space. If φ is a structural property, ψ is an aggregative property, and φ is sufficient for ψ then, if there is a proposition $P \in \wp(W)$ such that P has φ and $Pr(P) < 1$, it follows that $\emptyset$ has ψ.

Proof Since $Pr(P) < 1$, $P \neq W$ and, for some $w^* \in W$, $w^* \notin P$. For all $w_i \in W$, let π_i be a permutation on the elements of W such that $\pi_i(w_i) = w^*$, $\pi_i(w^*) = w_i$ and $\pi_i(w) = w$ for every other $w \in W$. Define $f_i(Q)$ as $\{\pi_i(w) \mid w \in Q\}$ for all $Q \in \wp(W)$. Each such f_i evidently meets the first two conditions for an automorphism. Each f_i also preserves the cardinality of propositions which, given that $\langle W, \wp(W), Pr\rangle$ is finite and uniform, ensures that it preserves the probability of propositions. In this case, each f_i is an automorphism of $\langle W, \wp(W), Pr\rangle$. Observe that, for each i, $w_i \notin f_i(P)$ (if $w_i \notin P$, then $P = f_i(P)$ and if $w_i \in P$, then $f_i(P)$ results from P by exchanging w_i and w*). Since, by stipulation, P has φ and φ is structural, it follows that, for all i,$1 \leq i \leq |W|$ $f_i(P)$ has φ and, thus, has ψ. Since ψ is aggregative, it follows that $f_1(P) \wedge \ldots \wedge f_{|W|}(P)$ has ψ, but $f_1(P) \wedge \ldots \wedge f_{|W|}(P) = \emptyset$. **QED**[3]

The significance of Impossibility 1 for refined rules should be clear. Let ψ be the property of justification and φ be a sufficient condition for justification as expressed by a refined rule. If multiple premise closure holds, it follows that ψ must be an aggregative property. If φ is structural and is satisfied by some proposition that is less than certain it follows, by Impossibility 1, that Ø will satisfy ψ. A broad range of potential refined rules turn out to express structural properties, including all of those defined in broadly logical and probabilistic terms.

A property φ of propositions was said to be structural just in case it is preserved by automorphisms. That is, a property φ of propositions is structural just in case, for any probability space $\langle W, F, Pr\rangle$, any proposition $P \in F$, and any automorphism f of $\langle W, F, Pr\rangle$, P has φ iff $f(P)$ has φ. It's clear that *having a certain probability value* or *having a probability value above a certain threshold* are structural properties in this sense. Say that an m-ary relation δ between propositions is structural just in case, for any probability space $\langle W, F, Pr\rangle$, any propositions $P_1 \ldots P_m \in F$ and any automorphism f of $\langle W, F, Pr\rangle$, $P_1 \ldots P_m$ stand in δ iff $f(P_1) \ldots f(P_m)$ stand in δ: $\delta(P_1 \ldots P_m)$ iff $\delta(f(P_1) \ldots f(P_m))$. It's clear that *having a greater probability than*, *being probabilistically independent of*, or *being a competitor proposition to* are structural relations in this sense.

[3] Douven and Williamson's proof also serves to establish the following, stronger theorem:

> **Impossibility 1*** Let $<W, \wp(W), Pr>$ be a finite, uniform probability space. If φ is a structural property with respect to $<W, \wp(W), Pr>$, ψ is an aggregative property and φ is sufficient for ψ then, if there is a proposition $P \in \wp(W)$ such that P has φ and $Pr(\varphi) < 1$, it follows that Ø has ψ.

Impossibility 1* is stronger than Impossibility 1 on account of the fact that any structural property will be structural with respect to $<W, \wp(W), Pr>$, but the converse need not hold. Impossibility 1 is, however, strong enough for Douven and Williamson's purposes.

It's also clear that Boolean combinations of structural properties and relations must themselves be structural: if δ is structural and, for any $P_1 \ldots P_m \in \mathsf{F}$, $\gamma(P_1 \ldots P_m)$ iff $\sim\delta(P_1 \ldots P_m)$ then γ must be structural. If δ is structural and γ is structural and, for any $P_1 \ldots P_m \in \mathsf{F}$, $\lambda(P_1 \ldots P_m)$ iff $\gamma(P_1 \ldots P_m) \vee \delta(P_1 \ldots P_n)$ (for $n \leq m$) then λ must be structural. Let δ be a structural binary relation between propositions. Let φ be a property of propositions such that, for any $P \in \mathsf{F}$, $\varphi(P)$ iff $\forall X \in \mathsf{F}\ \delta(P, X)$. According to this definition, a proposition possesses property φ just in case it stands in relation δ to all propositions in F. Since δ is structural, for any propositions P and $Q \in \mathsf{F}$ and automorphism f, $\delta(P, Q)$ iff $\delta(f(P), f(Q))$. It follows that $\forall X \in \mathsf{F}\ \delta(P, X)$ iff $\forall X \in \mathsf{F}\ \delta(f(P), f(X))$. Any automorphism is a 1:1 function from F onto itself—as such, for any proposition $Q \in \mathsf{F}$, there must be a proposition $R \in \mathsf{F}$, such that $Q = f(R)$. It follows that $\forall X \in \mathsf{F}\ \delta(f(P), f(X))$ iff $\forall X \in \mathsf{F}\ \delta(f(P), X)$. We have it that $\varphi(P)$ iff $\varphi(f(P))$ in which case φ is structural. This reasoning can clearly be generalised for properties of propositions defined in this way from structural m-ary relations using m – 1 quantifiers. Any property of propositions constructed from structural properties and relations using Boolean operators and first-order quantifiers will be structural.

We turn, finally, to higher-order quantification. The reasoning here is more convoluted, and may be harmlessly skipped, but is provided for the sake of thoroughness. Let $\mathcal{F}$ include the propositions in F, along with all the sets of propositions in F (first-order sets), sets of sets of propositions in F (second-order sets), sets of sets of sets of propositions in F (third-order sets) and (sets of)^i propositions in F (i-th-order sets), for any finite i. Let F^i be the set of i-th-order sets in $\mathcal{F}$. For any automorphism f, define the f-transform function f^* using the following recursive clauses:

(i) If $P \in \mathsf{F}$, then $f^*(P) = f(P)$
(ii) If $\Sigma \in (\mathcal{F} - \mathsf{F})$ then $f^*(\Sigma) = \{f^*(X) \mid X \in \Sigma\}$

The f-transform function, so defined, will be 1:1 function from $\mathcal{F}$ onto itself and, for any F^i, will impose a 1:1 function from F^i onto itself.

A relation δ between members of $\mathcal{F}$ might be described as structural just in case, for any probability space $\langle W, \mathsf{F}, Pr\rangle$, automorphism f of $\langle W, \mathsf{F}, Pr\rangle$ and $\Sigma_1 \ldots \Sigma_m$ in $\mathcal{F}$, $\Sigma_1 \ldots \Sigma_m$ stand in δ iff $f^*(\Sigma_1) \ldots f^*(\Sigma_m)$ stand in δ: $\delta(\Sigma_1 \ldots \Sigma_m)$ iff $\delta(f^*(\Sigma_1) \ldots f^*(\Sigma_m))$. Let δ be a structural binary relation between propositions in F and members of F^i for some i, and let φ be a property of propositions such that, for any $P \in \mathsf{F}$, $\varphi(P)$ iff

$\forall X \in F^i$, $\delta(P, X)$. According to this definition, a proposition possesses property φ just in case it stands in relation δ to all i-th-order sets. Since δ is structural, for any $P \in F$, $\Sigma \in F^i$ and automorphism f, $\delta(P, \Sigma)$ iff $\delta(f(P), f^*(\Sigma))$. It follows that $\forall X \in F^i$, $\delta(P, X)$ iff $\forall X \in F^i$, $\delta(f(P), f^*(X))$. Since f^* imposes a 1:1 function from F^i onto itself, there must be a $\Pi \in F^i$, such that $\Sigma = f^*(\Pi)$. It follows that $\forall X \in F^i$, $\delta(f(P), f^*(X))$ iff $\forall X \in F^i$, $\delta(f(P), X)$. We have it that $\varphi(P)$ iff $\varphi(f(P))$ in which case φ is structural. This reasoning can clearly be generalised for properties of propositions defined in this way from structural m-ary relations linking propositions and sets of varying order, using m – 1 quantifiers.[4] The upshot of the preceding results is that any property that can be constructed from basic probabilistic properties and relations using Boolean operators and first- and higher-order quantifiers will be structural. Any refined rule that, like Pollock's and Lehrer's, expresses such a property, will fall prey to Douven and Williamson's result.

Douven and Williamson advertise their argument as a generalisation of the various arguments that have been levelled at specific refined rules—such as the arguments against Pollock's and Lehrer's rules set out in section 9.1. There is something misleading about this, though. Douven and Williamson, as noted, rely upon strong assumptions about the underlying probability space—in particular they rely upon the assumptions that the underlying probability space is both finite and uniform. Without these assumptions, the proof of Impossibility 1 will not go through. Without these assumptions, there is no guarantee that the f_is, as defined in the proof of Impossibility 1, will be automorphisms of $\langle W, \wp(W), Pr \rangle$ in which case there is no guarantee that the $f_i(P)$s will share the structural properties of P. The arguments against Pollock's and Lehrer's rules, in contrast, rely on no such assumptions (see Chandler 2010). All that these arguments take for granted about the underlying probability space is that it be suitably *rich*—that it model enough lottery propositions to divide up any uncertain proposition into sufficiently unlikely subcases. There is, perhaps, something to be taken from these arguments that cannot be derived from Douven and Williamson's.

[4] As long as we restrict attention to finite probability spaces, this last result is strictly unnecessary. Relative to a finite probability space, any property of propositions that can be defined using higher-order quantifiers could also be defined using first-order quantifiers and Boolean operators. This result will assume some significance, though, when we come to consider infinite probability spaces, where this kind of translation is no longer guaranteed.

Both the finiteness and uniformity assumptions might indeed be thought questionable. If we take the 'possible worlds' talk at face value, then the finiteness assumption, at least, seems very much out of place. That is, if W is to be regarded as the totality of possible worlds and possible worlds are to be understood in the familiar way, then W will clearly be an infinite set. Douven and Williamson do suggest that the 'possible worlds' in W not be regarded as maximally specific—rather, W should be thought of as comprising a mutually exclusive and jointly exhaustive set of states that are specific enough to supply all possible answers to the questions that are relevant—in effect, each member of W will represent one possible combination of answers to the set of relevant questions (Douven and Williamson 2006: 775, 776). It is not entirely clear, though, that even this conception of the members of W will motivate the finiteness assumption—after all, certain questions permit of an infinite number of possible answers (such as those that can be answered with an arbitrarily high degree of precision). Neither, it should be pointed out, does this conception of the members of W provide any obvious motivation for the uniformity assumption. If anything, it makes it more difficult to see why such an assumption is warranted. After all, what entitles us to assume that each possible combination of answers to the set of relevant questions should be equally likely?

Even if Douven and Williamson's conception of the members of W succeeds in motivating the finiteness and uniformity assumptions, there is something ultimately dissatisfying about limiting our attention to such spaces. The full probabilistic bearing of a body of evidence is captured not by a finite probability space, delimited by a set of relevant questions, but by an infinite space in which *all* questions are deemed relevant, the members of W are deemed to be possible worlds, properly so speaking, and the set of propositions modelled is maximised. There is something to the thought that, if Douven and Williamson's result is unavailable for 'maximal' spaces such as these, then the project of devising refined rules in logical and probabilistic terms may have life in it yet.

It's important to note that Douven and Williamson do supply a proof of a related result that is not restricted to finite probability spaces. This is significant—but the result is, in some respects, weaker than Impossibility 1 and the proof continues to rely upon a fairly strong descendant of the uniformity assumption. I will undertake something similar here. That is, I will prove a slightly weakened version of Impossibility 1 that holds good for infinite probability spaces (see Smith 2010b). The weakening, though, is of a different kind—and a kind that is not, I think, significant. And the proof will not rely upon any uniformity-type assumption.

9.3. The Second Impossibility Result: Infinite Probability Spaces

The proof that I will outline in this section relies, like Douven and Williamson's, on some assumptions about the underlying probability space. It's worth noting that there is no prospect of escaping from such assumptions altogether—it's straightforward to show that there are some probability spaces for which Douven and Williamson's result will not hold. But the set of probability spaces consistent with my assumptions is, I think, of particular significance, for it plausibly includes within it the 'maximal' spaces, described in section 9.2, that represent the full probabilistic bearing of a body of evidence.

The first assumption I shall make is that of *countable additivity*. As mentioned in section 8.4, a probability function Pr will be countably additive iff it meets the following continuity condition: if P_i $i = 1, 2 \ldots$ is an increasing sequence of propositions ($P_1 \subseteq P_2 \subseteq P_3 \ldots$) then $Pr(\vee_i P_i) = \lim_{i \to \infty} Pr(P_i)$. If the domain of Pr is finite then this condition is automatically met. Countable additivity is a relatively standard constraint to impose once we allow for the possibility of infinite probability spaces—and it was a part of Kolmogorov's initial axiomatisation—but it is not uncontroversial and, thus, certainly worth noting.

Call a proposition Q a *sub-proposition* of P iff $Q \subseteq P$ and a *proper* sub-proposition of P iff $Q \subset P$. A proposition $P \in \mathsf{F}$ is said to be an *atom* of the probability space $\langle W, \mathsf{F}, Pr \rangle$ just in case $Pr(P) > 0$, and for all propositions $Q \in \mathsf{F}$, if Q is a proper sub-proposition of P, then $Pr(Q) = 0$. An atom is a proposition with positive probability, that has no proper sub-propositions with positive probability. If a probability space is finite then it must have atoms and, furthermore, every proposition that has positive probability will be equivalent to the disjunction of some atoms. In the kind of probability spaces that Douven and Williamson consider, the atoms are just the singleton sets containing the members of W.

If a probability space is infinite, however, then the possibility arises that the space be *atomless*. A probability space $\langle W, \mathsf{F}, Pr \rangle$ is said to be atomless just in case, for any proposition $P \in \mathsf{F}$ such that $Pr(P) > 0$, there is a proper sub-proposition Q of P, such that $Pr(P) > Pr(Q) > 0$. What atomlessness requires, in effect, is that any proposition with a positive probability has proper sub-propositions with lower positive probability. If $\langle W, \mathsf{F}, Pr\rangle$ is atomless, it follows that, for any $w \in W$, such that $\{w\} \in \mathsf{F}$, $Pr(\{w\}) = 0$. The second assumption I shall make is that of atomlessness.

The 'maximal' probability spaces that most faithfully capture the probabilistic bearing of evidence must, as discussed in section 9.2, be infinite. There is good reason to think that they must also be atomless. If the set of propositions modelled by a space is to be maximally rich then, for any proposition with a non-zero probability, it is plausible that we will always be able to identify some further probabilistically independent proposition that also has a non-zero probability. By conjoining the two, we will arrive at a proposition that is less likely than either conjunct, but has a probability greater than zero. Clearly, this could only be satisfied in an atomless probability space.

If the set W is *uncountably* infinite, then the simplifying assumption that the set of propositions F is equal to $\wp(W)$ becomes problematic, and we drop it here—if W is uncountably infinite then the claim that every subset of W receives a probability value is incompatible with certain constraints upon Pr. If a probability space is finite and uniform, then the propositions modelled in that space will receive only rational probability values. This follows straightforwardly from the observation made earlier—namely, that the probability of any proposition in a finite, uniform probability space will be equal to the ratio of the cardinalities of two finite sets. In an infinite probability space, it will be quite possible for propositions to receive irrational probability values. My proof, however, will continue to be limited to propositions that receive rational values—for reasons that will soon become evident. I will have more to say about this limitation in due course.

Let $\langle W, \mathsf{F}, Pr\rangle$ and $\langle W, \mathsf{F}', Pr'\rangle$ be two probability spaces such that $\mathsf{F}' \subset \mathsf{F}$ and Pr' is the restriction of Pr to the members of F'. Say, in this case, that $\langle W, \mathsf{F}, Pr\rangle$ is a *fine-graining* of $\langle W, \mathsf{F}', Pr'\rangle$ and $\langle W, \mathsf{F}', Pr'\rangle$ a *coarse-graining* of $\langle W, \mathsf{F}, Pr\rangle$. Fine-graining, in effect, augments the set of propositions modelled by a probability space while coarse-graining diminishes it. As I mentioned in section 9.2, whether a proposition possesses a property is, in general, something that is probability space relative—a proposition can possess a property relative to some probability spaces in which it features, but not others. Say that a property of propositions φ is *preserved by coarse-graining* just in case any proposition that possesses φ relative to a probability space must also possess φ relative to any coarse-graining of that space in which it features. More precisely, φ is preserved by coarse-graining just in case for any probability spaces $\langle W, F, Pr\rangle$ and $\langle W, F', Pr'\rangle$ such that $\langle W, F', Pr'\rangle$ is a coarse-graining of $\langle W, F, Pr\rangle$, and any proposition $P \in F'$, if P has φ relative to $\langle W, F, Pr\rangle$ then P has φ relative to $\langle W, F', Pr'\rangle$.

Many structural properties will be preserved by coarse-graining—the property of having a probability above a certain threshold is a simple example—but structuralness itself provides no guarantee of this.[5] As can be easily checked the conditions outlined in both Pollock's and Lehrer's rules are also properties that are preserved by coarse-graining. In fact, all of the extant refined rules that I am aware of share this feature. I think this is no accident. As I mentioned, all of these refined rules are specifically designed to exempt 'lottery propositions'. But lottery propositionhood, whatever it amounts to exactly, is some kind of extrinsic property that depends upon the availability in the space of further propositions with certain characteristics. Generally speaking, the more fine-grained a probability space, the easier it will be for a proposition to qualify as a lottery proposition and the more difficult it will be for a proposition to satisfy a refined rule. That is, if a proposition possesses the property expressed by a refined rule relative to a given probability space, then it will also possess this property relative to any coarse-graining of that space in which it features.

The result that I shall prove will be restricted to properties that are both structural and preserved by coarse-graining. It is in this way that it represents a weakening of Impossibility 1. A refined rule that expressed a structural property that is not preserved by coarse-graining would, in effect, evade my result. My suspicion is that seeking out rules of this kind would not prove fruitful—it is very difficult to see, at first blush anyway, what an independently motivated rule of this kind might look like. But,

[5] Consider the property of being *non-atomic*—that is, the property of having a proper sub-proposition with positive probability. As can be easily checked, this property is structural. Let $W = \{w_1, w_2, w_3\}$, $F = \wp(W)$ and Pr be a uniform distribution over the members of W. Let $F' = \{W, \{w_1, w_2\}, \{w_3\}, \emptyset\}$ and Pr′ be the restriction of Pr to F′. $\langle W, F', Pr'\rangle$ is a coarse-graining of $\langle W, F, Pr\rangle$, but $\{w_1, w_2\}$ is non-atomic with respect to $\langle W, F, Pr\rangle$ and atomic with respect to $\langle W, F', Pr'\rangle$.

beyond reporting this impression, I have little to say about such an undertaking. Douven and Williamson's result delivers a blow to the project of devising refined rules using just the resources of logic and probability theory—but, as discussed, it leaves open certain avenues by which this project might continue to be pursued. My aim here is, in effect, to block more of these avenues off. But there are some that yet remain open, in spite of what I will prove. This is one.

The proof I shall offer exploits a corollary of a result established by Villegas (1964) (see also Savage 1972: 37, 38)—a corollary to the effect that any proposition within an atomless probability space can always be partitioned into n equiprobable sub-propositions, for any positive integer n. What this shows is that, within an atomless probability space, there are always sufficient resources to construct a finite, uniform sub-space *around* a given proposition. This is the rough strategy that will be employed.

This construction will rely upon Zorn's Lemma. Let $(\Sigma, \leq)$ be a partially ordered set. A subset Π of Σ is described as a *chain* iff for all $x, y \in \Pi$, $x \leq y$ or $y \leq x$. The lemma states that, if Σ is a non-empty, partially ordered set, such that every chain in Σ has an upper bound, then Σ has a maximal element. Zorn's Lemma is, famously, set-theoretically equivalent to the axiom of choice.[6] I won't comment further upon its use here.

Before giving the proof, I shall introduce some further terminology. Let $\langle W, \mathsf{F}, Pr\rangle$ be a probability space with $\Omega \subset \mathsf{F}$ a finite and uniform partition of W—that is a set of n propositions $\{P_1, P_2, P_3, \ldots, P_n\}$ such that $Pr(P_1) = Pr(P_2) = Pr(P_3) \ldots = Pr(P_n) = 1/n$, $P_1 \vee P_2 \vee P_3 \vee \ldots \vee P_n = W$ and, for all positive integers i, j, $\leq$ n such that $i \neq j$, $Pr(P_i \wedge P_j) = 0$. Let $cl(\Omega) \subseteq \mathsf{F}$ be the closure of Ω under negation and disjunction. Call a function f a Ω-automorphism of $\langle W, \mathsf{F}, Pr\rangle$ just in case f is a 1:1 function from $cl(\Omega)$ onto itself that satisfies these conditions:

(i) $f(P \vee Q) = f(P) \vee f(Q)$
(ii) $\sim f(P) = f(\sim P)$
(iii) $Pr(P) = Pr(f(P))$
for all $P, Q \in cl(\Omega)$.

It is important to note that a Ω-automorphism of $\langle W, \mathsf{F}, Pr\rangle$ need only be partially defined upon F—its domain is $cl(\Omega) \subseteq \mathsf{F}$. Call a property φ of

6 That is, Zorn's Lemma and the axiom of choice are equivalent given the axioms of ZF.

propositions Ω-structural just in case, for any proposition $P \in cl(\Omega)$ and Ω-automorphism f, P has φ iff $f(P)$ has φ. All structural properties must be Ω-structural, for any Ω meeting the above conditions. This follows from the fact that $cl(\Omega)$ is itself closed under negation and disjunction and includes W, in which case all Ω-automorphisms of $\langle W, F, Pr\rangle$ will be automorphisms *simpliciter* relative to the coarse-graining $\langle W, cl(\Omega), Pr'\rangle$ (where Pr′ is the restriction of Pr to the members of $cl(\Omega)$). By the definition of a structural property, all structural properties must be preserved by all automorphisms of $\langle W, cl(\Omega), Pr'\rangle$. With this background in place, I shall prove the following:

Impossibility 2 Let $\langle W, F, Pr\rangle$ be a countably additive, atomless probability space. If φ is a structural property preserved by coarse-graining, ψ is an aggregative property and φ is sufficient for ψ then, if there is a proposition $P \in F$ such that P has φ relative to $\langle W, F, Pr\rangle$ and $Pr(P) = r/k$, for r and k positive integers with $r < k$, it follows that $\emptyset$ has ψ relative to some probability space.

Proof Let P be a proposition such that $Pr(P) = r/k$, for r, k positive integers with $r < k$. Call a proposition Q an *r-minor* sub-proposition of P just in case Q is a sub-proposition of P such that $Pr(Q) > 0$ and $Pr(Q) \leq Pr(P)/r$. By atomlessness, there is a decreasing sequence of sub-propositions of P, $Q_1, Q_2 \ldots$ such that for each Q_n, $Pr(Q_n) > 0$ and $\lim_{n\to\infty} Pr(Q_n) = 0$, in which case P is guaranteed to have an r-minor sub-proposition, for any positive integer r. Consider the set Π of all r-minor sub-propositions of P. This set can be partially ordered by entailment. If $R_1, R_2 \ldots$ is a chain of elements within this set (such that $R_1 \subseteq R_2 \subseteq \ldots$) then $\lim_{n\to\infty} Pr(R_n) \leq Pr(P)/r$ in which case, by countable additivity, $Pr(\vee_n R_n) \leq Pr(P)/r$. It follows that the union of the members of any chain of r-minor sub-propositions will itself be an r-minor sub-proposition and an upper bound to the chain. By Zorn's Lemma, then, the set of r-minor sub-propositions of P must have a maximal member. Let T be one such member.

Consider the proposition $P \wedge {\sim}T$. By atomlessness, there is a decreasing sequence of sub-propositions of $P \wedge {\sim}T$, $U_1, U_2 \ldots$ such that, for each U_n, $Pr(U_n) > 0$ and $\lim_{n\to\infty} Pr(U_n) = 0$. Since T is a maximal r-minor sub-proposition of P it follows that, for each n, $T \vee U_n$ is not an r-minor sub-proposition of P (because U_n is disjoint from T so

$Pr(T \vee U_n) = (Pr(T) + Pr(U_n)) > Pr(T))$; thus, for each n, $Pr(T \vee U_n) > Pr(P)/r$. So $\lim_{n\to\infty} Pr(T \vee U_n) \geq Pr(P)/r$. But $\lim_{n\to\infty} Pr(T \vee U_n) = \lim_{n\to\infty} (Pr(T) + Pr(U_n)) = Pr(T)$. Since T is an r-minor sub-proposition of P we have $Pr(P)/r \geq Pr(T)$. Thus, $Pr(T) \geq Pr(P)/r \geq Pr(T)$, in which case we have $Pr(T) = Pr(P)/r = 1/k$.

If $r = 2$ then $1/k = Pr(T) = Pr(P)/2 = (Pr(P \wedge T) + Pr(P \wedge \sim T))/2 = (Pr(T) + Pr(P \wedge \sim T))/2$, so $1/k = Pr(T) = Pr(P \wedge \sim T))$. If $r > 2$, we then seek out a maximal (r–1)-minor sub-proposition of $P \wedge \sim T$—call it V—which, by the above reasoning, will also have a probability of 1/k. If $r = 3$ then $Pr(V) = Pr(T) = Pr(P \wedge \sim T \wedge \sim V) = 1/k$. If $r > 3$, we seek out a maximal (r-2)-minor sub-proposition of $P \wedge \sim T \wedge \sim V$ and so on. After $r - 1$ repetitions of this process, P will be divided into r exclusive sub-propositions, each with a probability of 1/k. Proposition P will be equivalent to the disjunction of these r propositions. We then repeat the same process with respect to ~P, which, after $k - r - 1$ repetitions, will be divided into $k - r$ exclusive sub-propositions, each with a probability of 1/k. In this case W is divided into k equiprobable, exclusive, and exhaustive propositions. We have a uniform partition Ω of W of cardinality k such that $P \in cl(\Omega)$.

At this point, the proof, in essence, proceeds as before: since $Pr(P) < 1$, $P \neq W$ and, for some $Y^* \in \Omega$, Y^* is inconsistent with P. For all $Y_i \in \Omega$, let π_i be a permutation on the elements of Ω such that $\pi_i(Y_i) = Y^*$, $\pi_i(Y^*) = Y_i$ and $\pi_i(Y) = Y$ for every other $Y \in \Omega$. Define $f_i(Q)$ as $\vee\{\pi_i(Y) \mid Y \subseteq Q\}$ for all $Y \in \Omega$ and $Q \in cl(\Omega)$. Each such f_i evidently meets the first two conditions for a Ω-automorphism. Since the elements of Ω are equiprobable, it also meets the third condition in which case each f_i is a Ω-automorphism of $\langle W, F, Pr\rangle$. Since φ is preserved by coarse-graining, it follows that P has φ relative to $\langle W, cl(\Omega), Pr'\rangle$, where Pr' is the restriction of Pr to the members of $cl(\Omega)$. Since φ is structural, it follows that φ is Ω-structural and, for all i,$1 \leq i \leq k$ $f_i(P)$ has φ and, thus, has ψ. Since ψ is aggregative, it follows that $f_1(P) \wedge \ldots \wedge f_k(P)$ has ψ relative to $\langle W, cl(\Omega), Pr'\rangle$. But $f_1(P) \wedge \ldots \wedge f_k(P) = \emptyset$. **QED**[7]

[7] Interestingly, this proof does *not* suffice to establish a corresponding extension of Impossibility 1 as mentioned in n. 3:

> **Impossibility 2*** Let $\langle W, F, Pr\rangle$ be a countably additive, atomless probability space. If φ is a structural property with respect to $\langle W, F, Pr\rangle$ that is preserved by coarse-graining, ψ is an aggregative property and φ is sufficient for ψ then, if there is a proposition $P \in F$ such that P has φ and $Pr(P) = r/k$, for r

It is possible, then, to modify Impossibility 1 by adding the requirement that φ be preserved by coarse-graining and relaxing the requirement that $\langle W, \mathsf{F}, Pr \rangle$ be finite and uniform, allowing for the additional possibility that it be infinite and atomless (as well as countably additive).[8]

If we had begun by assuming that Pr(P) took an *irrational* value—one that could not be expressed as the ratio of two integers—then the proof would not have gone through. If Pr(P) were irrational, P could not be a member of cl(Ω), for any finite, equiprobable partition of W, Ω. There are, of course, structural properties, such that the only uncertain propositions that could possess them are those with irrational probability values. Seeking out a plausible refined rule that expresses such a property could perhaps be regarded as another avenue that remains open to the

and k positive integers with r < k, it follows that Ø has ψ relative to some probability space. From the assumption that φ is structural it will follow automatically that φ is structural with respect to $\langle W, cl(\Omega), Pr' \rangle$ (for $\Omega \subset \mathsf{F}$ and Ω a partition of W). But this will not follow from the weaker assumption that φ is structural with respect to $\langle W, \mathsf{F}, Pr \rangle$.

The following demonstration was pointed out to me by Stephan Leuenberger: let φ be the property of being atomic and *true*—that is, containing a designated 'actual' world. Since $\langle W, \mathsf{F}, Pr \rangle$ is an atomless space, φ will not be satisfied by any members of F and, thus, will count as trivially structural with respect to it. Since $\langle W, cl(\Omega), Pr' \rangle$ is atomic, φ will not be structural with respect to it, since truth is not preserved by automorphisms.

[8] I mentioned in n. 2 that Lin and Kelly (2012), Arló-Costa and Pedersen (2012), and Leitgeb (2014b) advance what are close to refined rules, featuring probabilistic properties of propositions that are closed under multiple premise deductive consequence. These properties are structural and thus, by Impossibility 1 and Impossibility 2, these rules trivialise in finite, uniform and infinite, atomless probability spaces (Lin and Kelly's 'camera shutter' rules are, in effect, ill defined for infinite, atomless spaces). None of these philosophers, however, take these rules to yield absolute verdicts about whether one has justification for believing a proposition—rather, these verdicts are taken to be *relative* to a given probability space or, as it might be put, relative to a given question (Lin and Kelly 2012: ss 13, 14) or a given partition (Leitgeb 2014b: s. 3). That is, none of these philosophers take these rules to articulate conditions under which one has justification *simpliciter* for believing a proposition—at best, they provide conditions under which one has justification for believing a proposition relative to a given representation of the probabilistic bearing of one's evidence.

When construed in this way, these rules might not be regarded as trivial, though their epistemic significance is somewhat unclear. There is of course nothing preventing us from introducing a technical notion of probability space relative justification and laying down some probabilistic conditions for it. Such a notion, though, is only liable to assume significance in a setting in which the very existence of justification *simpliciter* is being denied. And if the denial of justification *simpliciter* is simply motivated by the fact that we cannot identify non-trivial probabilistic conditions for it, then this denial is premature. I won't pursue this further here.

refined rule theorist—albeit another that would appear to hold little promise.

9.4. The Missing Ingredient

Technical results may never take us all the way to the philosophical conclusions we desire. There will always be a range of different lessons that one might choose to draw from results such as those established above. The lesson that I am inclined to draw, though, is this: the idea that we can have justification for believing things that are less than certain and the idea that justification is closed under relations of multiple premise deductive consequence cannot be reconciled using the tools of evidential probability alone. So long as we restrict ourselves solely to the resources of the risk minimisation conception, we will never be able to formulate a rule for justification that meets adequacy conditions (i) and (ii)—a rule that supplies justification for believing propositions that are less than certain, but does not supply justification for believing contradictions when combined with multiple premise closure. In order to turn this trick, some new ingredient is needed.

Consider the following:

Hybrid Normic Rule If $Pr(P) > t_1$ and $r(P) > t_2$ then one has justification for believing P.

where Pr is one's evidential probability function, r is one's *normic support* function, t_1 is a real number close to, but less than, 1, and t_2 is some non-negative integer. This rule will meet conditions (i) and (ii). Clearly, the property expressed by this rule can be satisfied by propositions that are less than certain. This property could not, however, be simultaneously satisfied by each member of an inconsistent set of propositions. By a result established in section 8.4, if a set of propositions is inconsistent then its members cannot each enjoy a positive degree of normic support. The rule can, then, be harmlessly combined with multiple premise closure—doing so will not supply justification for a contradiction.

The property described by this rule is not, of course, structural in Douven and Williamson's sense and, thus, escapes the above results. Douven and Williamson do acknowledge that rules describing non-structural properties may meet conditions (i) and (ii), but they give such rules relatively short shrift. With respect to rules expressing sufficient conditions for justification they state, 'The present paper will be concerned with . . . only those that are *formal* in the sense that they . . . are broadly probabilistic and logical,' and later, 'let us first say that in analytic philosophy the prima facie attractiveness of a formal approach should hardly need mentioning' (Douven and Williamson 2006: 758). If it is true at all that analytic philosophers ought to have a preference for formal rules, it is true only in some very broad sense of 'formal'—and certainly not in the rather loaded sense that Douven and Williamson have in mind.

The hybrid normic rule could perhaps be portrayed as a refinement of the low risk rule, its inclusion of a new sort of ingredient notwithstanding. Indeed, this rule could be made to conform to the template for a refined rule, with $r(P) \leq t_2$ serving as the exclusion condition C that is intended to root out lottery propositions. While it is possible to formulate rules that satisfy (i) and (ii) while staying relatively close to the low risk rule, there are other rules satisfying (i) and (ii) that depart from it more radically, such as the following:

Normic Rule If $r(P) > t$ then one has justification for believing P.

where r is one's normic support function and t is some non-negative integer. The pure normic rule evidently meets conditions (i) and (ii).

The pure normic rule represents, in a way, the normic equivalent of the low risk rule—it makes a certain degree of normic support sufficient for justification in the same way that the low risk rule makes a certain degree of evidential probability sufficient for justification. One could, perhaps, formulate more complicated rules of this kind—analogues of the various refinements of the low risk rule that have been proposed. In a way, though, I see no reason to tamper with the normic rule, as simple as it seems. Interestingly, appending Pollock or Lehrer-style exception conditions to the normic rule would prove superfluous—such exemptions are already implicit within the rule itself.[9]

[9] Assume that $r(P) > 0$ and that, for some proposition Q, $r(P \mid Q) < r(P)$. Applying the standard definition of conditional ranks, as set out in section 8.3, we have it that $r(Q \supset P) - r(\sim Q) < r(P)$. Since P entails $Q \supset P$ it follows that $r(Q \supset P) \geq r(P)$. Given both of these inequalities, we have it that $r(\sim Q) > 0$. But if $r(\sim Q) > 0$ it must be that $r(Q) = 0$ in which case $r(Q) < r(P)$. If $r(P) > 0$ then $r(P) > r(Q)$ for any proposition Q that is a competitor to P. This is closely connected to the *generalised rational monotonicity* principle for normic support

Any proposition P that is less than certain can be completely divided into subcases that are each as unlikely as its negation, if not more so. If I'm thinking in purely probabilistic terms, then believing P can seem a clumsy, ill-fitting attitude, no matter how close to certain P might be. After all, to believe P is to take a seemingly skewed or biased stance towards the possibilities before me—to single out the negation of P to dismiss, even though the other possibilities are, at best, on an epistemic par with it. To do this seems arbitrary.

If P is normically supported, however, then there is another perspective from which the subcases of P are not on par with its negation. The probability of a proposition P is like a limited resource that must be divided up as P itself is divided up—and the more subcases into which P is divided, the further this resource is stretched. Normalcy is not like this—not like a quantity that has to be 'shared out'. The subcases of P might all represent possibilities that are just as normal as P itself—and some of them, at any rate, *must do so*. If P is normically supported, then its negation represents an abnormal possibility while (some of) its subcases do not. When thinking normically, belief no longer seems such a peculiar attitude. Singling out the negation of P to dismiss, far from being arbitrary, now seems to be the attitude that most faithfully reflects my epistemic situation.

In *The Analogy of Religion* Butler famously quipped that, while an infinite intelligence deals only with certainties, 'for us, probability is the very guide to life' (Butler 1736: introduction). If the reflections of this book are along the right track then this isn't the right creed to adopt. And yet, my quarrel with Butler, and the many who have approvingly repeated his words since, need not, perhaps, be so profound. The guidance that probability can offer in our lives is of inestimable value—about this there is no quarrel. But, finite as we are, probability is not the only recourse that is available to us when certainty lies beyond our reach. In Chapter 2, I appropriated some words of Pietroski and Rey (1995: 84) intended originally to describe the contrast between ceteris paribus laws

proved in section 8.3. It is superfluous to add a Lehrer-style exemption condition to the pure normic rule. To see that it is also superfluous to add a Pollock-style exemption condition to the pure normic rule it suffices to observe, as noted above, that there can be no inconsistent set Σ such that, for each proposition $Q \in \Sigma$, $r(Q) > 0$. Thus, no proposition can be a member of such a set.

and brute statistical generalisations. To twist their words a little more, the notion of probability answers to the need to take gambles in an unpredictable world, while the notion of justification answers to the need to idealise in a complex world. Our world, of course, is both of these things.

Bibliography

Allensworth, R. (2009) 'Prediction markets and law: A skeptical account', *Harvard Law Review* 122(4), pp. 1217–29

Alston, W. (1985) 'Concepts of epistemic justification', *The Monist* 68(1), pp. 57–89

Alston, W. (1988) 'An internalist externalism', *Synthese* 74(3), pp. 265–83

Alston, W. (1995) 'How to think about reliability', *Philosophical Topics* 23(1), pp. 1–29

Arló-Costa, H. and Pedersen, A. (2012) 'Belief and probability: A general theory of probability cores', *International Journal of Approximate Reasoning* 53(3), pp. 293–315

Audi, R. (1993) *The Structure of Justification* (Cambridge: Cambridge University Press)

Audi, R. (1995) 'Memorial justification', *Philosophical Topics* 23(1), pp. 31–45

Audi, R. (2001) 'An internalist theory of normative grounds', *Philosophical Topics* 29(1/2), pp. 19–46

Barbey, A. and Sloman, S. (2007) 'Base rate respect: From ecological rationality to dual processes', *Behavioral and Brain Sciences* 30(3), pp. 241–97

Bar-Hillel, M. (1980) 'The base rate fallacy in probability judgments', *Acta Psychologica* 44(3), pp. 211–33

Bar-Hillel, M. and Wagenaar, W. (1991) 'The perception of randomness', *Advances in Applied Mathematics* 12(4), pp. 428–54

Bhaskar, R. (1975) *A Realist Theory of Science* (Leeds: Leeds Books Ltd)

Bird, A. (2007) 'Justified judging', *Philosophy and Phenomenological Research* 74(1), pp. 81–110

Blome-Tillmann, M. (2009) 'Contextualism, safety and epistemic relevance', *Philosophical Studies* 143(3), pp. 383–94

BonJour, L. (2010) 'The myth of knowledge', *Philosophical Perspectives* 24(1), pp. 57–83

Borges, J. (1935) *A Universal History of Iniquity*, Hurley, A. trans. (Penguin Books, 2004)

Boutilier, C. (1994) 'Conditional logics of normality: A modal approach', *Artificial Intelligence* 68(1), pp. 87–154

Boutilier, C. and Becher, V. (1995) 'Abduction as belief revision', *Artificial Intelligence* 77(1), pp. 43–94

Buchak, L. (2014) 'Belief, credence and norms', *Philosophical Studies* 169(2), pp. 285–311

Butler, J. (1736) *The Analogy of Religion* (London: J. M. Dent and Co., 1906)

Carnap, R. (1950) *The Logical Foundations of Probability* (Chicago: University of Chicago Press)

Chalmers, D. (2011) 'The nature of epistemic space', in Egan, A. and Weatherson, B. eds, *Epistemic Modality* (Oxford: Oxford University Press)

Chandler, J. (2010) 'The lottery paradox generalized?' *British Journal for the Philosophy of Science* 61(3), pp. 667–79

Chisholm, R. (1957) *Perceiving* (Ithaca, NY: Cornell University Press)

Christensen, D. (2004) *Putting Logic in its Place* (Oxford: Oxford University Press)

Cohen, L. (1977) *The Probable and the Provable* (Aldershot: Gregg Revivals)

Cohen, L. (1979) 'On the psychology of prediction: Whose is the fallacy?' *Cognition* 7(4), pp. 385–407

Cohen, S. (1984) 'Justification and truth', *Philosophical Studies* 46(3), pp. 279–95

Cohen, S. (1988) 'How to be a fallibilist', *Philosophical Perspectives* 2, pp. 91–123

Cohen, S. (1999) 'Contextualism, skepticism and the structure of reasons', *Philosophical Perspectives* 13, pp. 57–88

Comesaña, J. (2005) 'Unsafe knowledge', *Synthese* 146(3), pp. 395–404

Comesaña, J. (2009) 'What lottery problem for reliabilism?' *Pacific Philosophical Quarterly* 90(1), pp. 1–20

Comesaña, J. (2010a) 'An evidentialist reliabilism', *Noûs* 44(4), pp. 571–600

Comesaña, J. (2010b) 'Reliabilism', in Bernecker, S. and Pritchard, D. eds, *The Routledge Companion to Epistemology* (London: Routledge)

Conee, E. (2004) 'The truth connection' in Conee, E. and Feldman, R. eds, *Evidentialism: Essays in Epistemology* (Oxford: Oxford University Press)

Conee, E. and Feldman, R. (2004) 'Evidentialism', in *Evidentialism: Essays in Epistemology* (Oxford: Oxford University Press)

Conee, E. and Feldman, R. (2008) 'Evidence', in Smith, Q. ed., *Epistemology: New Essays* (Oxford: Oxford University Press)

Cosmides, L. and Tooby, J. (1996) 'Are humans good intuitive statisticians after all? Rethinking some conclusions from the literature on judgment under uncertainty', *Cognition* 58(1), pp. 1–73

Cournot, A. (1843) *Exposition de la théorie des chances et des probabilités* (Paris: Hachette)

Delgrande, J. (1987) 'A first order logic for prototypical properties', *Artificial Intelligence* 33(1), pp. 105–30

Derksen, A. (1978) 'The alleged lottery paradox resolved', *American Philosophical Quarterly* 15(1), pp. 67–74

DeRose, K. (1995) 'Solving the skeptical problem', *Philosophical Review* 104(1), pp. 1–52

DeRose, K. (1996) 'Knowledge, assertion and lotteries', *Australasian Journal of Philosophy* 74(4), pp. 568–80

DeRose, K. (2004) 'Sosa, safety, sensitivity and skeptical hypotheses', in Greco, J. ed., *Ernest Sosa and his Critics* (Oxford: Blackwell)

Descartes, R. (1628) *Rules for the Direction of the Mind* (Cambridge: Cambridge University Press, 1911)

Dougherty, T. (2011) 'In defence of propositionalism about evidence', in Dougherty, T. ed., *Evidentialism and its Discontents* (Oxford: Oxford University Press)

Douven, I. (2002) 'A new solution to the paradoxes of rational acceptability', *British Journal for the Philosophy of Science* 53(3), pp. 391–410

Douven, I. (2006) 'Assertion, knowledge and rational credibility', *Philosophical Review* 115(4), pp. 449–85

Douven, I (2008) 'The lottery paradox and our epistemic goal', *Pacific Philosophical Quarterly* 89(2), pp. 204–25

Douven, I. and Williamson, T. (2006) 'Generalizing the lottery paradox', *British Journal for the Philosophy of Science* 57(4), pp. 755–79

Dretske, F. (1970) 'Epistemic operators', *Journal of Philosophy* 67(24), pp. 1007–23

Dretske, F. (1971) 'Conclusive reasons', *Australasian Journal of Philosophy* 49(1), pp. 1–22

Earman, J. (1992) *Bayes or Bust? A Critical Examination of Bayesian Confirmation Theory* (Cambridge, MA: MIT Press)

Earman, J. and Roberts, J. (1999) '"Ceteris paribus" there is no problem of provisos', *Synthese* 118(3), pp. 439–78

Earman, J., Roberts, J., and Smith, S. (2002) 'Ceteris paribus lost', *Erkenntnis* 57(3), pp. 281–301

Easwaran, K. (2009) 'Probabilistic proofs and transferability', *Philosophia Mathematica* 17(3), pp. 341–62

Egan, A. (2006) 'Secondary qualities and self-location', *Philosophy and Phenomenological Research* 72(1), pp. 97–119

Enoch, D., Fisher, T., and Spectre, L. (2012) 'Statistical evidence, sensitivity and the legal value of knowledge', *Philosophy and Public Affairs* 40(3), pp. 197–224

Eriksson, L. and Hájek, A. (2007) 'What are degrees of belief?' *Studia Logica* 86(2), pp. 185–215

Falk, R. (1981) 'The perception of randomness', *Proceedings of the Fifth International Conference for the Psychology of Mathematics Education* (Grenoble: Laboratoire IMAG)

Falk, R. and Konold, C. (1997) 'Making sense of randomness: Implicit encoding as a basis for judgment', *Psychological Review* 104(2), pp. 301–18

Fallis, D. (1997) 'The epistemic status of probabilistic proof', *Journal of Philosophy* 94(4), pp. 165–86

Fantl, J. and McGrath, M. (2002) 'Evidence, pragmatics and justification', *Philosophical Review* 111(1), pp. 67–94

Fantl, J. and McGrath, M. (2009) *Knowledge in an Uncertain World* (Oxford: Oxford University Press)

Firth, R. (1978) 'Are epistemic concepts reducible to ethical concepts?', in Goldman, A. and Kim, J. eds, *Values and Morals* (Dordrecht: Reidel Publishing)

Foley, R. (1987) *The Theory of Epistemic Rationality* (Cambridge, MA: Harvard University Press)

Foley, R. (1992) 'The epistemology of belief and the epistemology of degrees of belief', *American Philosophical Quarterly* 29(2), pp. 111–21

Foley, R. (1993) *Working Without a Net* (Oxford: Oxford University Press)

Foley, R. (2009) 'Beliefs, degrees of belief and the Lockean thesis', in Huber, F. and Schmidt-Petri, C. eds, *Degrees of Belief* (Dordrecht: Springer)

Fumerton, R. (1995) *Metaepistemology and Skepticism* (Lanham, MD: Rowman and Littlefield)

Fumerton, R. (2004) 'Epistemic probability', *Philosophical Issues* 14(1), pp. 149–64

Goldman, A. (1976) 'Discrimination and perceptual knowledge', *Journal of Philosophy* 73(20), pp. 771–91

Goldman, A. (1979) 'What is justified belief?' in Pappas, G. ed., *Justification and Knowledge* (Dordrecht: Reidel)

Goldman, A. (2011) 'Towards a synthesis of reliabilism and evidentialism? Or: Evidentialism's troubles and reliabilism's rescue package', in Dougherty, T. ed., *Evidentialism and its Discontents* (Oxford: Oxford University Press)

Gundersen, L. (2004) 'Outline of a new semantics for counterfactuals', *Pacific Philosophical Quarterly* 85(1), pp. 1–20

Hájek, A. (2003) 'What conditional probability could not be', *Synthese* 137(3), pp. 273–323

Hambourger, R. (1980) 'Belief in miracles and Hume's Essay', *Noûs*, 14(4), pp. 587–604

Hammerton, M. (1973) 'A case of radical probability estimation', *Journal of Experimental Psychology* 101(2), pp. 242–54

Harman, G. (1968) 'Knowledge, inference and explanation', *American Philosophical Quarterly* 5(3), pp. 164–73

Harman, G. (1986) *Change in View* (Cambridge, MA: MIT Press)

Hawthorne, James (2009) 'The Lockean thesis and the logic of belief', in Huber, F. and Schmidt-Petri, C. eds, *Degrees of Belief* (Dordrecht: Springer)

Hawthorne, James and Bovens, L. (1999) 'The preface, the lottery and the logic of belief', *Mind* 108(2), pp. 241–64

Hawthorne, John (2003) *Knowledge and Lotteries* (Oxford: Oxford University Press)

Hazlett, A. (2006) 'How to defeat belief in the external world', *Pacific Philosophical Quarterly* 87(2), pp. 198–212

Hempel, C. (1965) *Aspects of Scientific Explanation and Other Essays in the Philosophy of Science* (New York: The Free Press)

Hempel, C. (1988) 'Provisos: A problem concerning the inferential function of scientific theories', *Erkenntnis* 28(2), pp. 147–64

Hill, C. and Schechter, J. (2007) 'Hawthorne's lottery puzzle and the nature of belief', *Philosophical Issues* 17(1), pp. 102–22

Huber, F. (2009) 'Belief and degrees of belief', in Huber, F. and Schmidt-Petri, C. eds, *Degrees of Belief* (Dordrecht: Springer)

Hume, D. (1748) *An Enquiry Concerning Human Understanding* (Oxford: Clarendon Press, 2000)

Ichikawa, J. (2014) 'Justification is potential knowledge', *Canadian Journal of Philosophy* 44(2), pp. 184–206

Jeffrey, R. (1956) 'Valuation and acceptance of scientific hypotheses', *Philosophy of Science* 23(3), pp. 237–46

Jenkins, C. (2007) 'Entitlement and rationality', *Synthese* 157(1), pp. 25–45

Jonakait, R. (1983) 'When blood is their argument: Probabilities in criminal cases, genetic markers and, once again, Bayes' theorem', *University of Illinois Law Review* 1983, pp. 369–421

Kahneman, D. and Tversky, A. (1972) 'On prediction and judgment', *Oregon Research Institute Research Bulletin* 12(4)

Kahneman, D. and Tversky, A. (1973) 'On the psychology of prediction', *Psychological Review* 80(4), pp. 237–51

Kaplan, M. (1995) 'Believing the improbable', *Philosophical Studies* 77(1), pp. 117–46

Kaplan, M. (2013) 'Coming to terms with our human fallibility: Christensen on the preface', *Philosophy and Phenomenological Research* 87(1), pp. 1–35

Kaye, D. (1982) 'The limits of the preponderance of the evidence standard: Justifiably naked statistical evidence and multiple causation', *American Bar Foundation Research Journal* 7(2), pp. 487–516

Keynes, J. (1921) *A Treatise on Probability* (London: Macmillan Press, 1973)

Klein, P. (1985) 'The virtues of inconsistency', *The Monist* 68(1), pp. 105–35

Klein, P. (1995) 'Skepticism and closure: Why the evil genius argument fails', *Philosophical Topics* 23(1), pp. 213–36

Koehler, J. (1996) 'The base rate fallacy reconsidered: Descriptive, normative and methodological challenges', *Behavioral and Brain Sciences* 19(1), pp. 1–17

Kolmogorov, A. (1933) *Foundations of Probability* (New York: Chelsea Publishing Company, 1950)

Korcz, K. (1997) 'Recent work on the basing relation', *American Philosophical Quarterly* 34(2), pp. 171–91

Kripke, S. (1980) *Naming and Necessity* (Oxford: Blackwell)

Kroedel, T. (2012) 'The lottery paradox, epistemic justification and permissibility', *Analysis* 72(1), pp. 57–60

Kvanvig, J. (2009) 'Assertion, knowledge and lotteries', in Pritchard, D. and Greenough, P. eds, *Williamson on Knowledge* (Oxford: Oxford University Press)

Kvanvig, J. and Menzel, C. (1990) 'The basic notion of justification', *Philosophical Studies* 59(3), pp. 235–61

Kyburg, H. (1961) *Probability and the Logic of Rational Belief* (Middleton: Wesleyan University Press)

Kyburg, H. (1970) 'Conjunctivitis', in Swain, M. ed., *Induction, Acceptance and Rational Belief* (Dordrecht: Reidel)

Kyburg, H. (1971) 'Epistemological probability', *Synthese* 23(2–3), pp. 309–29

Lehrer, K. (1974) *Knowledge* (Oxford: Clarendon Press)

Lehrer, K. (1980) 'Coherence and the racehorse paradox', in French, P., Uehling, T., and Wettstein, H. eds, *Midwest Studies in Philosophy* vol. 5 (Minneapolis: University of Minnesota Press)

Leitgeb, H. (2014a) 'A way out of the preface paradox', *Analysis* 74(1), pp. 11–15

Leitgeb, H. (2014b) 'The stability theory of belief', *Philosophical Review* 123(2), pp. 131–71

Lemos, N. (2007) *An Introduction to the Theory of Knowledge* (Cambridge: Cambridge University Press)

Leplin, J. (2009) *A Theory of Epistemic Justification* (Dordrecht: Springer)

Leslie, S. (2008) 'Generics: Cognition and acquisition', *Philosophical Review* 117(1), pp. 1–47

Levi, I. (1981) 'Should Bayesians sometimes neglect base rates?' *Behavioral and Brain Sciences* 4(3), pp. 342–3

Levi, I. (1983) 'Who commits the base rate fallacy?' *Behavioral and Brain Sciences* 6(3), pp. 502–6

Lewis, D. (1973) *Counterfactuals* (Oxford: Blackwell)

Lewis, D. (1979) 'Attitudes de dicto and de se', *Philosophical Review* 88(4), pp. 513–43

Lewis, D. (1980) 'A subjectivist's guide to objective chance', in *Studies in Inductive Logic and Probability* vol. 2 (University of California Press)

Lewis, D. (1996) 'Elusive knowledge', *Australasian Journal of Philosophy* 74(4), pp. 549–67

Lin, H. and Kelly, K. (2012) 'A geo-logical solution to the lottery paradox, with applications to nonmonotonic logic', *Synthese* 186(2), pp. 531–75

Littlejohn, C. (2012) *Justification and the Truth Connection* (Cambridge: Cambridge University Press)

Loftus, E. (1996) *Eyewitness Testimony* (Cambridge, MA: Harvard University Press)

Lopes, L. and Oden, G. (1987) 'Distinguishing between random and non-random events', *Journal of Experimental Psychology: Learning, Memory and Cognition* 13(3), pp. 392–400

Ludlow, P. (2005) 'Contextualism and the new linguistic turn in epistemology', in Preyer, G. and Peters, G. eds, *Contextualism in Philosophy* (Oxford: Oxford University Press)

Lycan, W. (2006) 'On the Gettier problem problem', in Hetherington, S. ed., *Epistemology Futures* (Oxford: Oxford University Press)

Lyon, D. and Slovic, P. (1976) 'Dominance of accuracy information and neglect of base rates in probability estimation', *Acta Psychologica* 40(4), pp. 287–98

McDowell, J. (1982) 'Criteria, defeasibility and knowledge', *Proceedings of the British Academy* 68, pp. 455–79

McGlynn, A. (2012) 'Justification as would-be knowledge', *Episteme* 9(4), pp. 361–76

McGlynn, A. (2014) *Knowledge First?* (Basingstoke: Palgrave Macmillan)

McHugh, C. (2014) 'Fitting belief', *Proceedings of the Aristotelian Society* 114(2), pp. 167–87

Makinson, D. (1965) 'The paradox of the preface', *Analysis* 25(6), pp. 205–7

Menzies, P. (2004) 'Difference making in context', in Collins, J. ed., *Causation and Counterfactuals* (Cambridge, MA: MIT Press)

Millikan, R. (1984) *Language, Thought and Other Biological Categories* (Cambridge, MA: MIT Press)

Moser, P. (1988) 'The foundations of epistemological probability', *Erkenntnis* 28(2), pp. 231–51

Nelkin, D. (2000) 'The lottery paradox, knowledge and rationality', *Philosophical Review* 109(3), pp. 373–409

Neta, R. and Rohrbaugh, G. (2004) 'Luminosity and the safety of knowledge', *Pacific Philosophical Quarterly* 85(4), pp. 396–406

Nickerson, R. (2002) 'The production and perception of randomness', *Psychological Review* 109(2), pp. 330–57

Niiniluoto, I. (1981) 'L J Cohen versus Bayesianism', *Behavioral and Brain Sciences* 4(3), pp. 349

Nozick, R. (1981) *Philosophical Explanations* (Cambridge, MA: Harvard University Press)

Owens, D. (2000) *Reason Without Freedom: The Problem of Epistemic Normativity* (London: Routledge)

Pappas, G. and Swain, M. (1978) 'Introduction', in Pappas, G. and Swain, M. eds, *Essays on Knowledge and Justification* (London: Cornell University Press)

Peijnenburg, J. and Atkinson, D. (2014) 'The need for justification', *Metaphilosophy* 45(2), pp. 201–10

Pietroski, P. and Rey, G. (1995) 'When other things aren't equal: Saving "ceteris paribus" laws from vacuity', *British Journal for the Philosophy of Science* 46(1), pp. 81–110

Plantinga, A. (1993) *Warrant and Proper Function* (Oxford: Oxford University Press)

Pollock, J. (1970) 'The structure of epistemic justification', in Rescher, N. ed., *Studies in the Theory of Knowledge* (Oxford: Blackwell)

Pollock, J. (1986) *Contemporary Theories of Knowledge* (Rowman and Littlefield)

Pollock, J. (1990) *Nomic Probability and the Foundations of Induction* (Oxford: Oxford University Press)

Priest, G. (2001) *An Introduction to Non-Classical Logic* (Cambridge: Cambridge University Press)

Pritchard, D. (2005) *Epistemic Luck* (Oxford: Clarendon Press)

Pritchard, D. (2007) 'Sensitivity, safety and anti-luck epistemology', in Greco, J. ed., *The Oxford Handbook of Scepticism* (Oxford: Oxford University Press)

Pritchard, D. (2012) *Epistemological Disjunctivism* (Oxford: Oxford University Press)

Pryor, J. (2000) 'The skeptic and the dogmatist', *Noûs* 34(4), pp. 517–49

Pryor, J. (2004) 'What's wrong with Moore's argument?' *Philosophical Issues* 14(1), pp. 349–78

Pryor, J. (2005) 'Is there immediate justification?' in Sosa, E. and Steup, M. eds, *Contemporary Debates in Epistemology* (Oxford: Blackwell)

Rabin, M. (1980) 'Probabilistic algorithm for testing primality', *Journal of Number Theory* 12(1), pp. 128–38

Ramsey, F. (1926) 'Truth and probability', in Braithwaite, R. ed., *Foundations of Mathematics and Other Essays* (London: Kegan, Paul, and Trubner, 1931)

Rantala, V. (1982) 'Impossible worlds semantics and logical omniscience', *Acta Philosophica Fennica* 35, pp. 106–15

Redmayne, M. (2008) 'Exploring the proof paradoxes', *Legal Theory* 14(4), pp. 281–309

Reed, B. (2010) 'A defence of stable invariantism', *Noûs* 44(2), pp. 224–44

Reynolds, S. (2013) 'Justification as the appearance of knowledge', *Philosophical Studies* 163(2), pp. 367–83

Ross, B. and Levy, N. (1958) 'Patterned prediction of chance events by children and adults', *Psychological Reports* 4, pp. 87–124

Russell, B. (1948) *Human Knowledge: Its Scope and Limits* (London: Unwin)

Ryan, S. (1991) 'The preface paradox', *Philosophical Studies* 64(3), pp. 293–307

Ryan, S. (1996) 'The epistemic virtues of consistency', *Synthese* 109(2), pp. 121–41

Savage, L. (1972) *Foundations of Statistics* (New York: Dover)

Schick, F. (1966) 'Consistency', *Philosophical Review* 75(4), pp. 467–95

Schiffer, S. (1991) 'Ceteris paribus laws', *Mind* 100(1), pp. 1–17

Schurz, G. (2001) 'What is "normal"? An evolution-theoretic foundation for normic laws and their relation to statistical normality', *Philosophy of Science* 68(4), pp. 476–97

Scriven, M. (1959) 'Truisms as grounds for historical explanations', in Gardiner, P. ed., *Theories of History* (New York: Free Press)

Señor, T. (1996) 'The prima/ultima facie justification distinction in epistemology', *Philosophy and Phenomenological Research* 56(3), pp. 551–66

Siegel, S. (2006) 'Which properties are represented in perception?' in Gendler Szabo, T. and Hawthorne, J. eds, *Perceptual Experience* (Oxford: Oxford University Press)

Siewert, C. (1998) *The Significance of Consciousness* (Princeton: Princeton University Press)

Skyrms, B. (1967) 'The explication of "X knows that p"', *Journal of Philosophy* 64(12), pp. 373–89

Smith, M. (2007) 'Ceteris paribus conditionals and comparative normalcy', *Journal of Philosophical Logic* 36(1), pp. 97–121

Smith, M. (2009) 'Transmission failure explained', *Philosophy and Phenomenological Research* 79(1), pp. 164–89

Smith, M. (2010a) 'What else justification could be', *Noûs* 44(1), pp. 10–31

Smith, M. (2010b) 'A generalised lottery paradox for infinite probability spaces', *British Journal for the Philosophy of Science* 61(4), pp. 821–31

Smith, M. (2012) 'Some thoughts on the JK-rule', *Noûs* 46(4), pp. 791–802

Smith, M. (2014) 'Knowledge, justification and normative coincidence', *Philosophy and Phenomenological Research* 89(2), pp. 273–95

Smithies, D. (2012) 'The normative role of knowledge', *Noûs* 46(2), pp. 265–88

Soames, S. (2002) *Beyond Rigidity: The Unfinished Semantic Agenda of Naming and Necessity* (Oxford: Oxford University Press)

Soames, S. (2011) 'Kripke on metaphysical and epistemic possibility', in Berger, A. ed., *Saul Kripke* (Cambridge: Cambridge University Press)

Sosa, E. (1999a) 'How to defeat opposition to Moore', *Philosophical Perspectives* 13, pp. 137–49

Sosa, E. (1999b) 'How must knowledge be modally related to what is known?' *Philosophical Topics* 26, pp. 373–84

Spohn, W. (1988) 'Conditional ordinal functions: A dynamic theory of epistemic states', in Harper, W. and Skyrms, B. eds, *Causation in Decision, Belief Change and Statistics* vol. 2 (Dordrecht: Kluwer)

Spohn, W. (2009) 'Survey of ranking theory', in Huber, F. and Schmidt-Petri, C. eds, *Degrees of Belief* (Dordrecht: Springer)

Spohn, W. (2012) *The Laws of Belief: Ranking Theory and its Philosophical Applications* (Oxford: Oxford University Press)

Stalnaker, R. (1976) 'Propositions', in MacKay, A. and Merrill, D. eds, *Issues in the Philosophy of Language* (New Haven: Yale University Press)

Stalnaker, R. (1978) 'Assertion', in Cole, P. ed., *Pragmatics* (New York: New York Academic Press)

Stalnaker, R. (1984) *Inquiry* (Cambridge, MA: MIT Press)

Stanley, J. (2005) *Knowledge and Practical Interests* (Oxford: Oxford University Press)

Stanley, J. (2008) 'Knowledge and certainty', *Philosophical Issues* 18, pp. 33–55

Stanovich, K. and West, R. (2000) 'Individual differences in reasoning: Implications for the rationality debate?' *Behavioral and Brain Sciences* 23(5), pp. 645–65

Steglich-Petersen, A. (2013) 'Truth as the aim of epistemic justification', in Chan, T. ed., *The Aim of Belief* (Oxford: Oxford University Press)

Sturgeon, S. (2008) 'Reason and the grain of belief', *Noûs* 42(1), pp. 139–65

Sutton, J. (2007) *Without Justification* (Cambridge, MA: MIT Press)

Swinburne, R. (2001) *Epistemic Justification* (Oxford: Oxford University Press)

Swinburne, R. (2011) 'Evidence', in Dougherty, T. ed., *Evidentialism and its Discontents* (Oxford: Oxford University Press)

Thompson, J. (1986) 'Liability and individualized evidence', in Parent, W. ed., *Rights, Restitution and Risk* (Cambridge, MA: Harvard University Press)

Villegas, C. (1964) 'On qualitative probability σ-algebras', *Annals of Mathematical Statistics* 35(4), pp. 1787–96

Vogel, J. (1990) 'Are there counterexamples to the closure principle?' in Roth, M. and Ross, G. eds, *Doubting: Contemporary Perspectives on Skepticism* (Dordrecht: Kluwer)

Vogel, J. (1999) 'The new relevant alternatives theory', *Philosophical Perspectives* 13, pp. 155–80

Weatherson, B. (2005) 'Can we do without pragmatic encroachment?' *Philosophical Perspectives* 19, pp. 417–43

Weatherson, B. (2011) 'Knowledge, bets and interests', in Brown, J. and Gerken, M. eds, *Knowledge Ascriptions* (Oxford: Oxford University Press)

Weatherson, B. (2014) 'Probability and scepticism', in Dodd, D. and Zardini, E. eds, *Contemporary Perspectives on Perceptual Justification and Scepticism* (Oxford: Oxford University Press)

Weatherson, B. (MS) 'Smith on justification and probability', <http://brian.weatherson.org/SoJaP.pdf>

Wedgwood, R. (2008) 'Contextualism about justified belief', *Philosophers' Imprint* 8(9), pp. 1–20

Wells, G. (1992) 'Naked statistical evidence of liability: Is subjective probability enough?' *Journal of Personality and Social Psychology* 62(5), pp. 739–52

Williamson, T. (1992) 'Inexact knowledge', *Mind* 101(2), pp. 217–42

Williamson, T. (2000) *Knowledge and its Limits* (Oxford: Oxford University Press)

Wright, C. (1992) *Truth and Objectivity* (Cambridge, MA: Harvard University Press)

Zalabardo, J. (2012) *Scepticism and Reliable Belief* (Oxford: Oxford University Press)

Index

Note: Page numbers in bold indicate the page at which the corresponding term is defined. A lowercase 'n' indicates a footnote on the associated page.

H

I

J

K

S

T

V

W

Z